Materials Handling

INDUSTRIAL ENGINEERING

A Series of Reference Books and Textbooks

Editor

WILBUR MEIER, JR.
Dean, College of Engineering
The Pennsylvania State University
University Park, Pennsylvania

Volume 1: Optimization Algorithms for Networks and Graphs, *Edward Minieka*

Volume 2: Operations Research Support Methodology, *edited by Albert G. Holzman*

Volume 3: MOST Work Measurement Systems, *Kjell B. Zandin*

Volume 4: Optimization of Systems Reliability, *Frank A. Tillman, Ching-Lai Hwang, and Way Kuo*

Volume 5: Managing Work-In-Process Inventory, *Kenneth Kivenko*

Volume 6: Mathematical Programming for Operations Researchers and Computer Scientists, *edited by Albert G. Holzman*

Volume 7: Practical Quality Management in the Chemical Process Industry, *Morton E. Bader*

Volume 8: Quality Assurance in Research and Development, *George W. Roberts*

Volume 9: Computer-Aided Facilities Planning, *H. Lee Hales*

Volume 10: Quality Control, Reliability, and Engineering Design, *Balbir S. Dhillon*

Volume 11: Engineering Maintenance Management, *Benjamin W. Niebel*

Volume 12: Manufacturing Planning: Key to Improving Industrial Productivity, *Kelvin F. Cross*

Volume 13: Microcomputer-Aided Maintenance Management, *Kishan Bagadia*

Volume 14: Integrating Productivity and Quality Management, *Johnson Aimie Edosomwan*

Volume 15: Materials Handling, *Robert M. Eastman*

Additional Volumes in Preparation

Materials Handling

Robert M. Eastman

Department of Industrial Engineering
University of Missouri
Columbia, Missouri

Marcel Dekker, Inc. New York • Basel

Library of Congress Cataloging-in-Publication Data

Eastman, Robert M.
Materials handling.

Includes index.
1. Materials handling. I. Title.
TS180.E18 1987 658.7'81 87-5277
ISBN 0-8247-7596-1

COPYRIGHT © 1987 MARCEL DEKKER, INC. ALL RIGHTS RESERVED

Neither this book nor any part may be reproduced or transmitted in any form or by any means, electronic or mechanical, including photocopying, microfilming, and recording, or by any information storage and retrieval system, without permission in writing from the publisher.

MARCEL DEKKER, INC.
270 Madison Avenue, New York, New York 10016

Current printing (last digit):
10 9 8 7 6 5 4 3 2 1

PRINTED IN THE UNITED STATES OF AMERICA

To Kathryn, Caroline, Parmelee, and Roger
for their unwavering support and help

Preface

Materials handling is moving product or supplies within a manufacturing or distribution establishment. According to best estimates, 20–25% of the labor dollar expended in American industry goes to materials handling. The percentage in distribution is higher. An activity that consumes resources of this magnitude is worth careful examination to maximize its efficiency at optimum cost.

Materials handling has gone through a technological revolution in the last few years. Exciting new developments include automatic guided (driverless) vehicles (AGV), automated storage and retrieval systems (AS/RS), flexible manufacturing systems (FMS), and robots. The essential component behind these developments is the modern digital computer. Its increase in capability, capacity, and speed combined with lower cost and smaller size have made technological improvements in materials handling technically and economically feasible. In this book we explore some of the new technologies and their application to modern materials handling systems.

The objective of the book is to furnish the reader with information on modern materials handling techniques, principles, equipment, and systems so that he or she can make intelligent decisions on materials handling system alternatives. It is not designed as a text in materials handling engineering for upper-level engineering students.

The audiences for which this book was written are:

1. Managers, supervisors, and staff whose positions call for decisions or advice on decisions regarding materials handling systems.

2. Noncredit courses and workships in materials handling. There is a real need here since many developments in recent years require that materials handling personnel update themselves in training, education, and experience to avoid technological obsolescence.
3. College courses which do not require a large design component. These programs include two- and four-year technology programs in industry, engineering, logistics, and related areas.

As mentioned above, the book is not designed to be a text for advanced engineering courses. It has been my experience and observation that the detailed design of equipment and system components is done by manufacturer engineering staffs, composed of engineers with formal training in the traditional fields of mechanical, electrical, and civil engineering. The formal training is supplemented by on-the-job and continuing education to provide the specialized knowledge needed and to update that knowledge periodically.

Most materials handling systems are built from standard compnents furnished by the manufacturer directly or through a consulting or contracting firm. Custom-designed materials handling system elements are exhorbitantly expensive and rarely justified economically. The system is designed by the manufacturer's staff or by the consulting firm to meet the customer's requirements.

The real need in the field of materials handling is knowledge by the user. Unless the user, whether manufacturer or distributor, understands the materials handling options available and the basic principles of materials handling system design, unsound decisions may result with large sums of money being spent for a system unsuited to the user's operations.

The concept for the book arose from my experience presenting noncredit materials handling workshops sponsored by the University of Missouri. This has given me valuable insight into the operating needs of many types of businesses and many examples and case studies to draw from. This has been supplemented by many years teaching credit courses in plant layout and materials handling at the University of Missouri.

The author assumes that the reader will have the following background:

1. Familiarity with manufacturing or distribution operations, especially those of his own organization. Broad experience is helpful but not essential. The author recommends that anyone whose duties include a major materials handling component join a professional organization such as the International Materials Management Society or the Institute of

Industrial Engineers, and attend conventions, shows, and open houses.

2. An interest in materials handling. Without interest and concern, motivation is lacking and little benefit will be gained.
3. Some knowledge of equipment operation and maintenance. Materials handling equipment is machinery with components and functions similar to those other types of machines. The care and maintenance of materials handling equipment are similar to those of other types of equipment. Familiarity with the basics of machinery and equipment will help with comprehension and application of the material presented in this book.

An author is inevitably faced with decisions on what to include and what to leave out. Including everything in materials handling would require an encyclopedia that might expand to several volumes. This is not my objective. My aim is to include the information and ideas needed for materials handling decision-making yet keep the size of the volume within reasonable limits. An encyclopedia is for reference on a specific topic; it is not a volume that covers a field in an organized, logical way for ease of comprehension.

Examples of areas that had to be restricted are robotics and safety in materials handling. Robotics is a new field with new material coming out regularly. Not only would it be difficult to cover the entire field of robotics construction, operation, and application, it is not feasible to present material that is completely up-to-date. Safety tends to be a specialized field with extensive literature of its own. Details of safety programs and equipment features are usually handled by specialists with safety engineering training and experience. In both these areas and in several others, the reader needs to understand some fundamentals but will normally call on specialists for detailed knowledge.

Coverage is broad rather than exhaustive. The objective is to help the materials handling manager and staff make informed choices among the multitude of materials handling system alternatives available. Detailed design and equipment choices require detailed investigation. In almost every case, choices will be among standard or nearly standard components combined into a system unique to the individual application.

Since this is a volume for the practitioner, extensive references are not included. It is assumed that the reader will be able to locate added information that he or she needs. This is not a research report; it is a volume to furnish practical help to the person faced with real materials handling problems.

I have discussed many aspects of materials handling with industrial engineering practitioners, alumni, and consultants. I was particularly interested to find out what approaches practitioners actually used in materials handling systems design and operation. For example, I found that computer programs were seldom used for facilities layout and system design. While some large companies with extensive resources made some use of computer facilities design programs, the overwhelming majority of firms made no use of them. Much of the work in computer-aided layout design is experimental research. It may be that in the future these programs will be efficient for general use. Today they are not. The aim of the book is to help the manager and staff with current information and practice. This is what we emphasize.

Also, I have used terms that are definite and precise yet are widely understood by the target audience. A materials handling design specialist may use other terms for specific situations.

Robert M. Eastman

Contents

Materials Handling

1

Introduction

1.1 INTRODUCTION

Materials handling is moving physical objects from one place to another. The material may be parts, components, subassemblies, raw materials, assemblies, or finished goods ready for shipment. The purpose of moving the material should be to increase its value.

Why is materials handling important to your organization? Money. Studies consistently show that materials handling costs average 20 to 25% of the manufacturing labor dollars in the United States. This estimate is probably low. For example, a production worker's material handling time may be charged as direct labor to a product even though this understates materials handling costs.

Materials handling costs vary widely among industries. In some, movement of materials may absorb a major proportion of the plant's labor dollar and require considerable investment in equipment. On the other hand, some specialized firms spend relatively little on moving materials. These usually have longer job cycles and high direct labor content in the product.

1.2 OBJECTIVE OF THIS BOOK

The objective of this book is to furnish the reader methods, techniques, and information for solving materials handling problems in his organization. We'll first discuss identifying the specific problem and possible courses of action; second, possible solutions including various types of materials handling equipment; third, evaluating the alternatives and making a choice or recommendation; and finally, implementing the chosen alternative.

The reader may have one or more different roles. He or she may be the staff engineer or technician who gathers data on the problem and information on equipment alternatives. The engineer may be asked to identify and define the problem and propose solutions. The person may be the production supervisor looking for ways to cut costs and improve operations or the executive who makes the decision on materials handling alternatives. A reader may fill more than one of these roles.

1.3 APPROACH OF THIS BOOK

This book is for the reader who is actively engaged in operating or designing materials handling systems and who is looking for practical help for his problems. It assumes that the reader knows his industry and is familiar with common industrial terminology and practice. It doesn't assume that the reader has had specialized or advanced training in materials handling or systems analysis.

1.4 THE SYSTEMS APPROACH

In this book, we shall take the systems approach. This means that we will look at materials handling as an organized movement pattern which is connected with other functions in the factory. In the systems approach we consider all affected activities and facilities as well as the materials handling system itself. The goal of the systems approach is the most efficient operation for the entire factory, rather than efficient but uncoordinated subsystems which may not add up to the best overall result.

1.5 DETERMINING THE OBJECTIVES

The first step you as a material handling engineer will undertake is determining the objectives of the materials handling system and the constraints imposed on the project. The goals of the layout will determine its major features as well as many mincr features. For example, the major goal of the system may be increased production to keep up with growing demand. In another situation, it may be to reduce costs to remain competitive.

1.6 DEFINING THE PROBLEM

Clear definition of the materials handling problem is essential to a successful solution. Time and effort spent on solving the wrong problem is wasted. A clear, unambiguous problem statement will facilitate reaching a good solution efficiently.

One way to identify the real problem is to ask, "What are we really trying to accomplish?" It may be to reduce labor cost, improve productivity, relieve a bottleneck, or eliminate a safety hazard. If feasible, the problem should be stated in numerical terms.

Constraints are limitations on the alternatives the materials handling engineer must observe in his system design. The most common is fitting the system into an existing building. Management may specify a maximum capital expenditure on the project. A deadline for completion may be imposed.

In any situations, goals and constraints may not be explicitly stated to the layout engineer. Through judicious questioning and observation, the engineer should find out what management hopes to attain with the new layout and act accordingly. One of the most frustrating events in a young engineer's career is to be told, "It's a fine layout. We can't use it because management won't let us spend that much money."

Some constraints are imposed by government. For example, fire codes require provisions for access and egress, sprinkler systems, segregation of flammable materials, and fire retardant construction. Sanitary officials will require sanitary and personal facilities for the employees and prohibit features deemed undesirable. Zoning may limit the type of building, its size, landscaping, parking areas, placement on the lot, and many other features. Building codes specify what materials and construction features may or may not be used. Environmental regulations specify limits on the release of pollutant materials into the atmosphere or ground water systems.

1.7 DATA

Once the materials handling engineer has determined the goals, objectives, and constraints, the next step is gathering data on which to base the system design. Much of this information is fuzzy and indeterminate. This is particularly true of sales and production volume estimates. Some data may be obtained from existing records; other data may require special observation. The time available for data gathering is limited. The engineer often must use a less-than-optimum amount of data and proceed with his systems design in order to meet management's deadlines.

1.8 MATERIALS HANDLING EQUIPMENT

Materials handling equipment comes in a wide range of types, models, makes, and sizes. It ranges from simple hand trucks to sophisticated robots and automatic transfer lines. The variety can be bewildering. However, this variety can be classified into categories which can be examined in groups of similar machines or equipment types. The characteristics, applications, advantages, and disadvantages of each category will be discussed.

1.9 EVALUATING A MATERIALS HANDLING SYSTEM

After a materials handling system has been designed, it should be evaluated by several persons from different points of view. The evaluation judges the system for its efficiency in meeting goals and objectives, for its economy, for its technical and operational suitability, and for errors or omissions that must be corrected.

The original evaluation is made by the system designer himself. However, a basic engineering principle is that at least one person other then the designer should check and evaluate the system design. Usually several people review the materials handling system.

The plant engineer reviews the plan for proposed building and equipment acquisitions or alterations and for maintenance requirements. Labor relations looks at the job descriptions and staffing. Line supervision is interested in its impact on production. Safety and legal departments may be involved. The financial and tax departments look at the economic justification.

Finally, the responsible management official makes the final decision approving, disapproving, or modifying the proposal. Only after all evaluations are completed, modifications made, and executive approval received does the project implementation go ahead.

1.10 IMPLEMENTATION

After the system design is finished and drawn up, the job is not done. The best materials handling system design in the country is valueless unless it is accepted and implemented by management. A layout seldom sells itself. The engineer must sell it to his superiors. This requires effective packaging, careful preparation, and convincing presentation.

Once management has approved the layout, implementation starts. This is usually supervised by the plant engineering department using either company personnel or outside contract workers. The materials handling engineer is called on for consultation during the implementation phase to make adjustments for unforeseen difficulties, management directives, or changing conditions. A materials handling system should be regarded as dynamic rather than static. The best system will require constant updating and improvement as conditions dictate modifications.

The final step in implementation is the follow-up. The engineer should review the installation periodically during the change phase and for a while thereafter. Inevitably there will be problems unforeseen in the original system design, and few persons are in a better position to advise on changes needed.

1.11 MATERIALS HANDLING DESIGN ORGANIZATION

The materials handling engineer designs and implements systems of moving materials through the factory. In a large organization, materials handling design may be assigned to specialized individuals or a separate department. In a small one, materials handling may be combined with other job functions. Materials handling is often regarded as a separate function because of the variety of materials handling equipment, because materials handling may be a different labor catagory, and because of the amount of money involved.

1.12 SERVICE MATERIALS AND FACILITIES

A factory needs indirect materials and supplies which must be moved even though they support production and don't become part of the product. These include lubricants, spare parts, tools and fixtures, housekeeping supplies, fuel, and office supplies. The plant must also have facilities for service functions, such as plant engineering, tool cribs for the storage, issue, and maintenance of small tools, sanitation, food, and time keeping. Utilities must be provided; these affect the design of the materials handling system. Scrap and waste must be removed from the production floor and the factory.

1.13 RECEIVING AND SHIPPING

A major part of the overall materials handling system is receiving and shipping. Areas must be provided for incoming trucks and rail cars, loading docks, storage of raw materials and finished goods, and for personnel and paperwork involved.

1.14 NONMANUFACTURING ORGANIZATIONS

We will not limit our discussion of materials handling to manufacturing industries. Other organizations need good materials handling to operate efficiently. A distribution warehouse must have provision not only for efficient receiving, shipping, and storage, but methods for assembling various items into shipments to customers. A hospital must handle quantities of linen, laundry, drugs, equipment, and housekeeping supplies. Even a large office organization moves paperwork, stationery, and supplies. Good materials handling principles apply no matter what the product or service involved.

2

Defining the Materials Handling Problem

2.1 INTRODUCTION

Before the materials handling engineer can design an improved system, he must define the problem. Good problem definition is important; care and effort here will pay off in better solutions with less cost and effort. A poor or incorrect problem statement can produce a poor solution or even wasted effort.

The steps to a good problem definition are:

1. Establish the subject of the problem.
2. Define the scope.
3. Define the objectives.
4. List the constraints.
5. Determine the outcomes expected.
6. Organize the method of analysis and solution.

2.2 WHAT IS THE PROBLEM?

2.2.1 What Is the Assignment?

Usually, the materials handling system designer is given a specific problem assignment by the executive in charge. This states what the boss wants done. Either at the beginning or later, the designer may have to request a clarification of vague or unspecified points or a change in the problem owing to changed conditions. In any case, the designer should keep the assignment in mind and concentrate all efforts toward carrying it out efficiently and promptly.

It is important to define the problem that the materials handling system is to solve before the design work starts. Too often a project is started before the person doing the work is fully informed of the problems that led to the decision to install a new system and of the goals and objectives that the organization and its managers are striving to reach.

The most important question to answer is "What is the organization trying to accomplish with the new system?" To increase production? To reduce costs? Is it moving to a new location to abandon a facility that is outmoded, obsolete, and impossible to expand because of surrounding development? The materials handling system design engineer should endeavor to find out, directly or indirectly, the answer to the first question above before he starts his job.

A good problem assignment should follow the well-known principles of a journalist's story. It should answer the questions: What? Who? Where? When? Why? and How? Typical questions are:

1. What materials are to be moved? What equipment is available? What are we trying to accomplish?
2. Who is going to do the moving? Who is affected by the new materials handling system? Who wants the work done? Who is in charge? Who has to approve?
3. Where is the material being moved from and to? Where does the system fit into other factory operations?
4. When must material be moved? How fast? When must the project be finished and in operation?
5. Why is the change being made? What are the reasons for instituting the change?
6. How will the material be moved? Unit loads? Packaging? Equipment? Power sources?

Another way of looking at the goals and objectives is to look at the factors that brought about the problem and the decision in the first place. Some of the possible answers are:

1. Organization goals. The company may be entering a new market which requires a new manufacturing facility. It may be expanding to meet increased demand due to shifting buying habits or to increased population.
2. Push by the chief executive officer. The chief executive officer may be pushing for a new layout or new facility as part of his personal program which may differ from the program and goals of the organization.
3. A critical incident may have occurred. A critical incident is one that brings about a drastic change in direction or action by

the company. These could include government action on environmental pollution, costs and prices that have become noncompetitive, and problems in obtaining enough labor of the right skills.

4. Competition is a strong motivating factor in many decisions. It is axiomatic in business that if you don't move forward, the competition will. Then you will fall behind and lose out. A steady state in industry does not exist.

2.2.2 Methods of Determining the Subject

Sometimes the subject of the materials handling assignment is incomplete, unspecified, or left to the designer to determine and refine. The designer should review the assignment and make a personal visit to the area for better understanding.

Observation

An essential step to gaining information and insight is a personal visit to the area of observation. This will clarify the situation, the subject, the equipment, the present methods of materials handling, the limitations, and other important factors. Often observations will disclose unknown obstacles which must be surmounted or, conversely, opportunities for improvement not noticed previously.

Checklists

A designer can sometimes use a checklist profitably. A typical, useful one has been included in *Basics of Material Handling* by Raymond A. Kulwiec, published by the Material Handling Institute, Inc. (1, p. 20).

MATERIAL HANDLING CHECKLIST

Is the material handling equipment more than 10 years old?

Do you use a wide variety of makes and models which require a high spare-parts inventory?

Are equipment breakdowns the result of poor preventive maintenance?

Do the lift trucks go too far for servicing?

Are there excessive employee accidents due to manual handling of materials?

Are materials weighing more than 50 pounds handled manually?

Are there many handling tasks that require two or more employees?

Are skilled employees wasting time handling materials?

Does material become congested at any point?

Is production work delayed owing to poorly scheduled delivery and removal of materials?

Is high storage space being wasted?

Are high demurrage charges experienced?

Is material being damaged during handling?

Do shop trucks operate empty more than 20% of the time?

Does the plant have an excessive number of rehandling points?

Is power equipment used on jobs that could be handled by gravity?

Are too many pieces of equipment being used, because their scope of activity is confined?

Are many handling operations unnecessary?

Are single pieces being handled where unit loads could be used?

Are floors and ramps dirty and in need of repair?

Is handling equipment being overloaded?

Is there unnecessary transfer of material from one container to another?

Are inadequate storage areas hampering efficient scheduling of movement?

Is it difficult to analyze the system because there is no detailed flow chart?

Are indirect labor costs too high?

Checklists should be used with care and understanding. If the list is too long, there's a tendency to run through the list without really giving thought to each item. If too short, the list may not include a point that is vital to the study.

Symptoms

A symptom is a "subjective evidence of disease or a physical disturbance" or "something which indicates the existence of something else" (2, p. 1196). A symptom of poor materials handling indicates that the system is deficient. It does not necessarily tell the designer what is wrong and thus point directly to the solution. It is easy to confuse the symptom with the problem. For example, the executive may direct, "Clean up the bottleneck in the receiving department! " but this doesn't identify the problem, which may be inadequate space, poor scheduling of incoming shipments, or cumbersome procedures.

Basics of Material Handling has a list of common symptoms of materials handling problems which may be useful to the system designer (1, p. 4).

SYMPTOMS OF INEFFICIENT MATERIAL HANDLING

Backtracking in material-flow path
Built-in hindrances to flow
Cluttered aisles
Confusion at the dock
Disorganized storage
Excess scrap
Execssive handling of individual pieces
Excessive manual effort
Excessive walking
Failure to use gravity
Fragmented operations
High indirect labor costs
Idle machines
Inefficient use of skilled labor
Lack of cube storage
Lack of parts and supplies
Long hauls
Material piled up on the floor
No standardization
Overcrowding
Poor housekeeping
Poor inventory control
Product damage
Repetitive handling
Service areas not conveniently located
Trucks delayed or tied up
Two-man lifting jobs

2.2.3 Relationships

An important part of any problem definition is the relationship of the system being studied to other components of the factory operation. Where does the incoming material come from? In what condition? In what type of container? In what quantity? Where does it go when the system discharges it? Shipping? Another production department? Storage?

Also, how does the system interface with other materials handling systems in the factory? Will interchangeability, compatibility, and flexibility be required? Can they be achieved?

2.3 DEFINING THE SCOPE

2.3.1 Bounding the Materials Handling System

In order to keep the materials handling system design project manageable, it must be bounded by establishing outside limits beyond which the project will not go. Although the goal of optimizing the operations of the factory with a single project would be ideal, its scope is simply too great for efficient analysis and solution. However, bounding the project does not mean ignoring the interfaces and interactions with other components of the factory.

A boundary may be set almost anywhere. The following principles will help the manager and designer set logical and efficient boundaries which will aid rather than inhibit the solution:

1. Physical boundaries such as a plant or a wall may be used.
2. A department.
3. An entire moving conveyor even though it may include more than one department.
4. A boundary separating the system from another system which can be efficiently handled as a different project.

2.3.2 What's Given?

The executive may lay down several specifications for the task. These become the requirements of the system design, but may be modified later as further data are obtained and as the project develops.

Product

The first element given is the product to be moved. This may be a single component or a mixture of several products. It should be accompanied by a complete set of drawings and specifications.

Space

The second element to be specified is the space involved. What area? How big is it? How high is the ceiling? How much must be available for aisles and fire lanes? Are there any obstructions overhead?—these are just a few of the space questions that should be answered.

Equipment

What equipment is involved in the project? This category could include:

1. Machinery processing the product
2. Heating, ventilating, and air conditioning

3. Present material handling equipment
4. Auxiliary equipment such as hoists, cranes, lift trucks
5. Lockers, water fountains, and similar personnel service equipment
6. Utilities, such as power

Personnel

The assignment should state which personnel are affected. These include:

1. Materials handling workers
2. Production workers
3. Supervisors
4. Plant engineering
5. Production and inventory control
6. Quality assurance
7. Receiving and shipping

The assignment should also indicate the extent to which these people should be involved and consulted.

Time

The time parameters should also be determined. The two major catagories here are the date the materials handling system should be in operation and the amount of personnel and support time available for the project.

Resources

The assignment should contain authorization for expending the firm's resources on the project. Some important points are:

1. What professional and support personnel will be available to work on the project? When? For how long?
2. How much money can be spent on the materials handling system project design?
3. How much money can be spent on purchasing, installing, and implementing the new materials handling system?

2.3.3 Planning Horizon

The planning horizon is defined as the number of future years covered by the company's planning. This horizon will differ for different expenditures and functions. For example, planning for a new facility may cover a period of several years. Planning for the number of items to be produced may cover only a week or a month.

The planning horizons in a typical company are:

Planning horizon	Function	Purchasing
1 month	Production schedule for the factory	For immediate requirements
3 months	Production schedule with some flexibility	For those items which must be committed within a short time in order to secure delivery in 3 months
1 year	Annual model design and specifications, total annual production and sales estimates	Tooling and short-run equipment
3 years	Design planning for complex assemblies such as automibiles and appliances	Specialized production equipment for high-volume, long-run production; equipment with service life of 1–5 years
5 years	This is the limit of definitive planning for new equipment, products, processes, and facilities Broad outlines with enough flexibility to change as conditions change	
10 years	General plans only	
Over 10 years	Rather rare; there are too many uncertainties to justify any but the most general goal statements and long-range plans	

2.3.4 What's Left Out?

After determining what's included, the designer should take a few moments to observe what's left out. Key equipment and obstructions may have inadvertently been left out of the instructions. The executive may have deliberately left something out of the assignment because of impending product or process changes which have not been announced or some other reason he does not want to divulge.

2.4 OBJECTIVES

An objective is "something toward which effort is directed" (4, p. 814). Unless the objective of a materials handling project is clearly defined and understood, there is significant likelihood that the result will be less than satisfactory. One way to identify objectives is to ask the question "What are we trying to accomplish?"

Objectives may be open or hidden. A manager may state the objectives of the project clearly and unambiguously. If so, good. However, he may have an objective that is omitted or left unstated. He may want to create more positions in his department to enhance his importance and justify a higher salary for himself. Or he may want to restructure the work to get rid of a troublesome employee.

There can be more than one objective of the materials handling system. In some cases, multiple objectives are complementary or noninterfering. Increased productivity and lower costs may well go together. An objective may conflict with another. For example, providing better customer service and reducing the inventory are mutually opposing objectives and are difficult to accomplish simultaneously.

Let's look at possible objectives of a material handling project. These will vary in importance from project to project.

2.4.1 To Reduce Costs

Cost reduction is an ever-present concern of operating management. Trends toward higher costs abound. They include inflation, escalating labor wages plus benefits, and inertia. Only by controlling costs in all phases of the business can the organization remain profitable and stay in business.

Before we can design the system to reduce costs, we must know what costs are being considered. Among the possibilities are the following:

Total Materials Handling Cost

This includes all costs that can be assigned directly to the materials handling function. This category includes materials handling labor plus fringe benefits and overhead, materials handling equipment costs, and materials consumed in material handling. Here the company is striving to optimize its materials handling function by minimizing the sum of its costs.

Materials Handling Labor Cost

This objective may be satisfactory if labor is a large component of materials handling cost. It has the advantage of being simple to identify and aggregate. It assumes that if materials handling labor costs are minimized, the total materials handling cost will be minimized also.

Inventory Costs

Every dollar of inventory costs the company about 30 cents per year before income taxes for interest, handling, storage, obsolescence, and other charges. Even if only return on investment is considered, this is currently about 20% per year of the inventory value.

Other Costs

This category includes damage, maintenance, lost orders, and expediting expenses due to a poor system. These are usually corollary to other cost objectives rather than being a primary objective.

2.4.2 To Improve Customer Service

In a competitive economic system, customer service is a potent weapon in the struggle for sales and profits. The customer wants and demands on-time deliveries, accurate order filling, specified quality, minimum damage, and prompt attention to special requests. If he cannot get this from your company, he will purchase elsewhere.

Faster Customer Service

A prime objective of a materials handling system is shorter delivery cycle from receipt of the order to delivery to the customer. A good materials handling system will accomplish this through faster processing throughput and shorter manufacturing cycle times.

More Efficient Shipping and Receiving

A common bottleneck in manufacturing is the shipping and receiving operation. It often receives less attention since it is regarded as a nonproductive overhead department. Shipping and receiving should be regarded as an integral part of the manufacturing system.

Reduce Damage and Delay

A good materials handling system will cut down on the damage in handling and moving materials and product. It will also reduce the amount of material lost or mislaid in the factory. One large manufacturer reported that a major materials handling problem was material lost or misidentified right on the shop floor.

2.4.3 To Reduce Work-in-Process Inventory

In this era of high interest and carrying charges, management is under strong financial pressure to reduce inventories. A good materials handling system can help accomplish this by moving material through production faster, by avoiding large banks of work awaiting processing, and by closer coordination of sales and inventory.

2.4.4 To Increase Productivity

Increasing productivity of the factory may be another way of stating objectives of reducing cost and improving throughput. If the manufacturing system can increase the output in relation to the resources used, sales and profits should both rise. This objective may be stated to increase output per man-hour or per square foot of floor space or for the total factory system.

2.4.5 To Improve System Control

Although this objective may not be the main one, it is an important subobjective and by-product of good material handling system design. A good system enables the production and inventory control department to locate and keep track of specific orders and to take any action necessary to meet schedules and shipping dates.

2.4.6 Better Personnel Utilization

The company may want to reduce the number of workers assigned to materials handling job classifications. This will reduce costs and at the same time reduce the number of long-term commitments to employees. The company may have trouble getting qualified materials handling workers. Materials handling wages may be too high relative to the worth of the output. The work may be hard and undesirable, leading to high turnover. It may be necessary to reduce heavy lifting.

2.5 CONSTRAINTS

The materials handling system designer needs to establish the constraints on his project at the start. Perhaps these may be modified later as pertinent evidence is gathered. Some are specific to the project; others commonly apply to all projects. The latter group includes:

1. Area. The department, function, product, or area of the plant needs to be specified. Optimizing the materials handling system of the entire factory may be theoretically desirable, but it is usually more feasible to limit the scope of a project to a smaller area or subsystem.

   At the same time, the system designer must take into account those related functions which interface his project and make sure his design does not adversely affect some other area of the factory.
2. Money. Most layout projects have a ceiling on cost or a requirement of a specified rate of return. A standard pay-off period or some other limitation may be placed on the money that can be spent.
3. Time. The layout engineer is normally given a deadline by which the proposed layout work must be completed. Often this deadline is too close to permit detailed analysis of all possible alternatives. A layout engineer must use his skills, experience, and judgment to decide how much time can be devoted to each phase of the project.
4. Personnel. The layout engineer may be limited in the number and skills of the people assigned to his layout project. He also has to consider the availability of supporting services such as design, drafting, and duplicating. Another personnel limitation is the number and type of workers to be hired to operate the new plant. What skills are required? Can these skills be obtained locally? What work must be subcontracted and done elsewhere because of lack of equipment, money, or skills?

2.6 EXPECTED OUTCOMES

Usually a target outcome for a materials handling system design project is specified or implied in the assignment. The designer should keep the targets in mind during the study and strive to reach those targets with his solution.

2.6.1 Operational Capabilities

The most frequent outcome expected by management is the ability of the materials handling system to reach specified operating rates. These may be in units per hour or day, dollar volume of product per unit time, tonnage per day, or some other measure of production output per unit time. Often the source of the materials handling system improvement project is the need to increase the productive capacity of the plant or department.

Other operational characteristics may be laid down. These might include a maximum damage level, accident prevention, or noise reduction. Another is compatibility with existing product lines, packaging, unit loads, and equipment.

2.6.2 Reports

Usually the results and recommendations of the study are presented as a report. Company standard operating procedure may define the format and content of the report. In other situations, the project engineer may have some leeway in organizing his report.

How Formal and Elaborate a Report?

This is determined by a variety of factors. The desires and working style of the manager directing the study, the amount of money involved, the availability of support personnel for report graphics and reproduction, all influence the extent of the report. Some executives want the report to be comprehensive and all-inclusive. Others want a brief report which is little more than a summary with supporting data available, if requested. Admiral Ernest King, Navy Chief of Staff during World War II, wanted short reports and returned any longer than one page for rewriting and abridging.

Drawings and Specifications

Again, the instructions and desires of the executive in charge will determine the engineering drawings and specifications to be included. A common practice is for the materials handling system designer to specify the broad outlines of the system and have engineers, detailers, or draftsmen in the plant engineer's department prepare the drawings and specifications for equipment procurement and installation.

2.6.3 Extent of Recommendations

Some executives want the materials handling system designer to limit his efforts and recommendations to the problem assigned. Others look for initiative in recommending improvements that may not have been envisioned in the original report. Again, this is usually a question of the executive's working style and has to be learned by observation, judicious questioning, or informal means.

2.7 METHOD OF ANALYSIS AND SOLUTION

The final step in defining the problem is designing the method of analysis and the approach to the solution. This will help refine the problem statement and determine whether the problem can be readily solved. The method of analysis and solution may point out changes in the problem definition that will improve the analysis, reduce the time and effort required, and produce a better solution.

Of great importance is determining the data needed and its source. Some may be easily available through product and plant drawings and specifications. Other data may require observation and study on the shop floor. Some data may not be available at all or else be so difficult and expensive to obtain that another route must be chosen.

In the next chapter, we shall discuss the data that may be needed and the sources for obtaining it.

REFERENCES

1. R. A. Kulwiec, *Basics of Material Handling*, The Material Handling Institute, Inc., Pittsburgh (1981).
2. By permission. From Webster's Ninth New Collegiate Dictionary © 1986 by Merriam-Webster, Inc., publisher of Merriam-Webster® Dictionaries.

3

Data for Materials Handling System Design

3.1 INTRODUCTION

Good data are essential to designing an efficient materials handling system on schedule. We define data as information, preferably quantitative, on the characteristics of the product, the production processes, and the physical facilities.

The data-gathering phase of the project will include collecting information, opinions, and requests on many factors from different sources. The value of the project in money invested and money saved will determine the amount and detail of data. It is uneconomical to spend a large sum of money improving an operation that involves only a limited volume of production and minor value.

This chapter contains a brief description of the data needed for designing a materials handling system. The forms used will be described in more detail in Chapter 6 along with some examples.

3.2 SOURCES OF DATA

Data can be obtained from several sources. The materials handling system designer must determine which data are needed and the best source for getting that data accurately and quickly. He should also verify that the data are what he needs and that there are no errors or biases, accidental or deliberate, in the information. Among the sources of materials handling data are the following:

3.2.1 Manufacturer's Data Sheets

These furnish data on the capacities, size measurements, speeds, clearances, engineering specifications, and other characteristics of equipment being considered for installation.

3.2.2 Production and Inventory Control Department

This group can provide data on volume of production, fluctuations, inventory requirements, and work-in-process cycles.

3.2.3 Personal Observation

The materials handling engineer may have to obtain some data through personal observation and measurement. Building obstructions, condition, and characteristics should be checked in person. The architect's drawings may not have been updated for alterations and modifications. Actual construction may have deviated from the original plans. Data that the materials handling engineer needs may not have been important to the architect who drew up the original plans.

Some data come from physical measurement on the scene. Aisle and building clearances should be measured and checked. Some movement times may not be available from the work measurement department. There may be some question about the suitability of existing data for materials handling system design.

In any event, the materials handling engineer should visit the site personally as frequently as he can. All too often, mistakes creep into a design that could be avoided by personal observation on the factory floor.

3.2.4 Handbooks

Much useful data can come from a handbook of engineering data. Standard aisle widths, rail car clearances, truck turning radii, vehicle parking requirements, angle of repose for bulk materials, beam strengths, and deflections are just a few of the data available in standard handbooks.

3.2.5 Engineering Design Department

Data on product characteristics will come from the engineering design department. A set of drawings and specifications will usually provide the needed information. Dimensions, weights, shapes, and special characteristics of the product are examples.

If a large number of different products are involved, drawings and specifications of a few typical products which are representative and which contribute the largest volume should suffice. The engineering department should be able to provide information on pending changes in the product that will affect the materials handling system design.

3.2.6 Plant Engineering

The plant engineering department is responsible for the physical facilities of the factory and handles maintenance of the building, utilities, machinery, and equipment. Plant engineering personnel perform or supervise the installation of new materials handling system equipment. They should be able to provide building and equipment data essential to the design of the materials handling system.

3.2.7 Sales Department

The sales department should have information on impending product changes, expected trends in sales volume, product mix, and other factors affecting the materials handling system. This is especially important in planning for future expansion and alteration.

3.2.8 Analogies

Sometimes, there are no solid data based on engineering design or operating experience. If this is the case, the materials handling engineer will usually design by analogy to similar existing systems. The results should be carefully checked to ensure that the analogy is appropriate and that the design meets engineering requirements for performance, safety, and other criteria.

3.3 WHAT IS TO BE MOVED?

The first determination is "What is to be moved?" It may be a single product, a family of similar products, or a wide variety of items. Among the product characteristics the materials handling engineer needs to know are the following:

3.3.1 Dimensions

The length, width, and height of the product must be determined. Also, can the product be nested? An example is automobile fenders, which can be stacked so that a number of fenders will take much less space than they would if not nested. The dimensions of a standard unit load must be determined if the product is so handled.

3.3.2 Weight

The weight of each item handled is important to good materials handling system design. Some products are light enough that weight is no consideration; the weight of others requires careful design consideration.

3.3.3 Sanitation

In some industries, such as food processing, meat packing, and beverages, sanitation is vitally important to the manufacture of a salable product.

3.3.4 Safety

The materials handling system must move the product without endangering the workers. It must prevent the worker from becoming entangled in the conveyors or being struck by falling objects. Protruding parts must be avoided. Radioactive or toxic materials and products required special precautions.

3.3.5 Fragility

The system must move the material expeditiously while minimizing damage and breakage during transportation. Fragile items require special handling which must be designed into the system.

3.3.6 Other

In addition to data on size, shape, weight, special requirements, and vulnerability to damage, the designer needs other information. The product mix is important; this is the proportion of each different product in the production stream. Future product plans are important. A good system is flexible to accommodate new products, requirements, and equipment.

3.4 WHERE DOES THE PRODUCT MOVE FROM AND GO TO?

The materials handling engineer must know where the material moves. In some cases, the layout may be set and the system designer must tailor his recommendations to the existing layout. In the more fortunate situation, the materials handling design is an integral part of the original factory layout design. This will usually result in a more efficient layout both from a facilities and from a materials handling point of view.

The designer needs a layout of the existing system. This may be available from existing drawings. If so, the work is reduced. It may have to be drawn from other drawings or from measurements made on the shop floor. Often, drawings for one purpose are incomplete or unsuitable for a different one. The detail and format may vary; it is uneconomical to draft elaborate layouts when a simple drawing or sketch will do.

The location of the beginning and end points of the product movements must be determined. This may appear obvious. Sometimes, the apparent answer is not the most desirable one and a change will improve the overall productivity.

A recent trend is to have work-in-process inventory in continuous motion until it is used. An example is an assembly plant for large appliances which carries about 1 day's inventory of work-in-process components on an overhead conveyor until the parts are required on the assembly floor.

A layout drawing can be an elaborate, complete engineering drawing or a simple sketch. The latter is usually enough for the materials handling system designer and is quicker and cheaper. The complete drawing may be advisable for presentation to top management and may be required by company standard operating procedure for engineering purposes.

3.4.1 Charts

The operation process chart is used to show how the product is assembled and the order in which parts and subassemblies are incorporated into the product. The assembly operation chart can be used to work out the production flow in the factory. The starting point is the assumption that the ideal flow in the factory will follow the order of parts processing and assembly into the final product.

The assembly operation chart can be expanded to show the times for each operation, the storages, and the distances moved. The author's preference is to keep the assembly operation chart simple. If more detail is desired, it can be put into supporting tables and charts.

If the product is complex with a large number of parts and subassemblies, a complete chart could cover a large wall and be so large and complex that it is hard to use. A better practice is to restrict the complexity of a chart by showing subassemblies and even individual components on separate charts which are referenced in the main chart.

3.5 WHY IS THE MATERIAL BEING MOVED?

This question may seem unnecessary. However, it will sometimes disclose moves that should not be made at all. Elimination of useless movements may be the biggest source of savings in a new materials handling system.

A flow process chart is a good way to examine the steps in the production, processing, or distribution of a product. The engineer can examine and question each step with emphasis on moves. Often, the chart will disclose unnecessary moves which can be eliminated or ones which could be shortened or modified to increase productivity.

A route sheet is a listed sequence of operations necessary to manufacture the product or component or to perform the task. It will list the equipment used, the operations performed at each work station, and other information such as tooling. Some data will be necessary to the materials handling engineer. Other information is not needed for the materials handling problems but is necessary to a functioning production system. A manufacturer will usually have route sheets for the products made in the factory.

3.6 WHO WILL MOVE THE MATERIAL?

This merits careful consideration since an unskilled or semiskilled materials handling worker can easily cost $10 an hour in basic wage, $4 an hour in direct fringe benefits, and $6 an hour in overhead, for a total cost of $20 an hour to the employer (1982). One of the major objectives of a good materials handling system is to minimize the number of workers required. The upward trend in wages and benefits will probably exceed the rate of increase in the cost of equipment.

Often production workers are required to move the material to and from their work station. This is usually uneconomical and tends to disrupt the even flow of production. If at all possible, material should be moved mechanically and automatically or by materials handling workers. In some unionized factories, materials handling work is limited to those whose job classification call for materials handling duties.

The flow process chart illustrated later can be used to determine who does what materials handling. It can identify materials handling being done by production, maintenance, or skilled workers.

The cheapest materials handling may be no manual intervention at all. This eliminates the wage cost of materials handling almost completely. However, an automated materials handling system requires a high production volume to be economical. If production volume is too low, fluctuates too much, or is interrupted frequently, a highly mechanized materials handling system may prove expensive and uneconomical.

The collective bargaining agreement may limit the changes in job assignment possibilities. The number and availability of professional, technical, and support personnel may constrain the scope of the project. The availability of skilled workers for installation of the system may influence what can be accomplished.

3.7 WHEN SHOULD MATERIAL BE MOVED?

Again this question looks obvious. It should be moved as fast as possible through the plant. This reduces the work-in-process inventory and improves the plant productivity. However, there are several considerations to be examined.

1. Is the materials handling continuous or intermittent? Some factories, such as an automobile assembly plant, will move the automobile continuously through the assembly line once it starts into production. On the other hand, an auto accessory factory will move spare-parts orders intermittently in small lots.
2. Is the factory on 24-hr operation? This is essential in some industries in which the machinery and processes cannot be shut down. However, this poses problems such as "When will the maintenance be done?" and "What happens when emergency repairs are needed on the graveyard shift when the number of experienced, qualified maintenance workers on duty is minimal?"

3.8 HOW MUCH MATERIAL MUST BE MOVED?

The volume of product moved is a key factor in materials handling system design. A separate discussion is devoted to volume determination and considerations. A few important observations are:

1. What is the total volume of product to be moved per unit time?
2. What are the fluctuations in volume? Are there daily, weekly, seasonal variations in the amount of product produced?
3. What provision should be made for expansion of the materials handling system as the production volume increases?

3.9 WHAT ARE THE PRESENT LAYOUT AND EQUIPMENT?

The system designer will need existing drawings and specifications on the present system, the building and its fixtures, the production equipment, and the product. He or she should verify that these are accurate and up-to-date. Too often modifications are made without the corresponding drawing change.

3.10 SUMMARY

In this chapter we have discussed the data that the materials handling system designer will need. This may vary depending on the extent of the investment involved. It's up to the materials handling system designer to collect the data needed to complete the work without expending time and effort on data that is marginally useful or even unnecessary. A rule of thumb is that professional employee time will cost the employer 0.1% of the annual salary for each hour expended. For example, a $30,000-a-year engineer will cost the company about $30 per hour. This covers salary, direct fringe benefits, and overhead. The designer should ask himself, "Is what I'm doing on this project worth what my time costs my employer?"

4

Materials Handling Equipment

4.1 INTRODUCTION

In this chapter, we'll look at the types of materials handling equipment available to the systems designer. In each category, there are many makes, models, and variations. Representative items are illustrated.

Materials handling systems are normally put together from standard components and equipment. Custom equipment is rarely justified from a benefit/cost viewpoint. It is easier, quicker, and cheaper to use a standard item as manufactured. If the standard item won't do, it is usually possible to modify it to suit the application.

There are many manufacturers of equipment both in the United States and abroad. Some sell direct to the ultimate user. Others operate through a manufacturer's representative or distributor. The manufacturer must provide for spare parts and maintenance service no matter which distribution channel is chosen. The purchaser should verify availability of parts and service since this is vital to satisfactory operation of a materials handling system.

In recent years, there has been little change in the basic mechanical and electrical design of materials handling equipment. There have been improvements in reliability, safety, strength, and controls. The most important change in materials handling is the sophisticated control and operation brought about by the tremendous improvement in computer capabilities and by the reduction in computer cost and size.

4.2 CRITERIA FOR EQUIPMENT SELECTION

A criterion is a "a standard upon which a judgment or decision can be based" (1, p. 307). There are two different categories of criteria for choosing equipment. The first consists of those used to decide which type of equipment is best for the materials handling system. The second includes the ones used to select a specific make and model after the general type of equipment for the system has been established.

4.2.1 Cost

The most important factor in equipment decisions is cost. The company wants to minimize the cost of performing the materials handling function consistent with satisfactory performance. This is essential if the company is to maintain and enhance its profit and economic position.

Investment

The first cost element is the capital investment required to acquire the asset. According to current accounting practice, this includes the following:

1. The invoice price
2. Freight from the source to the purchaser's plant
3. Costs of unloading, millwright work, and installation
4. Test and run-in expenses

Another definition of capital investment is the sum of those costs incurred until the equipment is ready to produce on a regular basis.

Investment is particularly important since any amount spent for materials handling equipment is unavailable for any other purpose. Every organization has requests and opportunities for capital investment that add up to much more than the total funds available. Management must exercise care to avoid tying up capital funds in assets that do not contribute to the financial health of the company.

On the other hand, total reliance on the capital investment criterion is unwise. The lowest capital cost alternative may have high operating expenses, unsatisfactory operating characteristics, high maintenance expenses, or poor reliability.

Operating Expense

Operating expenses are those costs incurred in regularly using the equipment. In accounting terms, they are the expenditures that are not treated as capital investment but are written off during the period in which they were incurred. They include labor, direct fringe benefits, fuel and energy, maintenance, personal property taxes, insurance, spare parts, and other current items.

A significant proportion of operating costs goes to labor and direct fringe benefits. New materials handling systems and equipment should decrease the labor required for both direct operation and maintenance. Some companies require that major cost-saving investments show reduction in number of workers as well as in dollars. In one large auto company, securing authorization to add a permanent position to the personnel roster is more difficult than securing approval for a capital investment.

4.2.2 Reliability and Maintainability

As the pace of production in a modern factory rises, the cost of downtime and interruptions also rises. Equipment should be chosen for its ease of maintenance (maintainability) and for its reliability (minimum downtime). Data for this criterion are not easy to obtain. In addition, equipment performance in these characteristics is dependent in part on the user's maintenance and operating policies. If the company does only minimum emergency repair and no preventive maintenance, the reliability and maintainability will be inferior to that of equipment well cared for in another company.

4.2.3 Service Facilities

The availability, quality, and accessibility of service facilities are important. The owner of equipment should be able to secure spare parts and good maintenance service quickly and cheaply from the manufacturer or his representative. The owner should perform basic preventive maintenance and many small repairs with his own personnel. However, there are limits to what the owner can do; he must have good service when needed.

4.2.4 Operating Characteristics

Another important factor in equipment selection is the operating characteristics of the equipment. These include speeds, capacities, acceleration and deceleration, and weight.

Many of these are zero-one criteria. The equipment either meets the specification or it doesn't. On the other hand, characteristics exceeding the specified requirements are unnecessary and may be unnecessarily expensive.

4.2.5 Safety and Environmental Characteristics

In recent years, these have become more important because of the increased regulation of occupational health and safety and of environmental pollution. The recommended equipment should meet or exceed minimum requirements. In some cases, features exceeding the minimum might be worthwhile.

Since materials handling accounts for a substantial proportion of industrial accidents, materials handling systems including the equipment should be designed and chosen to keep accidents to a minimum. An accident is expensive not only in direct costs but in the much larger amount lost from indirect costs of lost time, temporary replacements, confusion, and damaged property.

4.2.6 Compatibility and the Systems Concept

The systems concept calls for each component to interface efficiently with the rest of the system. This requires that the individual items of equipment be compatible with the rest of the system and its equipment. For example, all pallets within the system should be the same size. Different sizes will complicate handling and storage and lead to confusion and delays.

4.3 CONVEYORS

A conveyor is a mechanism that moves material along a fixed path. It may move continuously or intermittently. A conveyor may transport material to, through, and from receiving, shipping, processing, assembly, or storage. It may be operated by electric, gravity, or manual power.

Webster defines a conveyor as "a mechanical device for carrying packages or bulk material from place to place (as by an endless moving belt or a chain of receptacles)" (1, p. 287). It can be continuous or intermittent, general-purpose or specialized, fixed position or portable. Conveyors are widely used to move materials efficiently and cheaply since they do not require a manual worker during movement. Some require a worker to load and unload the product; others use automatic devices to load and unload.

4.3.1 Wheels and Rollers

A common, efficient type of conveyor moves the product along on rotating wheels or rollers. A powered or "live" roller conveyor can move packages up a 7° slope as well as horizontally. If the rollers are unpowered, the product moves by gravity or manual effort. A wheel conveyor is shown in Figure 4.1.

Wheel and roller conveyor systems are usually designed and installed on a custom basis with little provision for modification or

Fig. 4.1 Wheel conveyor. (Courtesy Conveyor Systems, Inc.)

flexibility. The salvage or scrap value of a conveyor is usually zero or less. However, portable conveyors are available for applications requiring shifting the conveyor from one spot to another. An example is piling incoming coal for outside yard storage.

An example of a roller conveyor is shown in Figure 4.2. Rollers can be power-driven or unpowered. If the material is moved by gravity or by pushing, power is not necessary. Power is needed to move the product up an incline or along a horizontal conveyor at a constant rate. If the conveyor is powered, provision must be made to avoid accidents, pileups, or damage.

Roller conveyors can be designed to fit the building and production operations. They can also be built to be easily moved as the production and layout needs change.

Fig. 4.2 Roller conveyor. (Courtesy Conveyor Systems, Inc.)

4.3.2 Fixtures, Slats, and Bars

Many factories use a floor-mounted chain conveyor to move heavy, bulky materials. Special attachments can be used to hold and position the product as it moves along the line.

The most spectacular example of a specialized in-floor conveyor is the automobile assembly line. The chassis is placed in a fixture on the line which moves at a steady rate. Components and parts are added as it moves. Figure 4.3 shows an automobile assembly line.

4.3.3 Chutes, Slides, and Belts

A chute is an inclined surface with sides for materials movement by gravity. It is simple and cheap to install and uses cost-free power. Common applications include bulk material such as coal and grain or simple transfer of packages where no sortation or power is required.

A moving belt is a relatively inexpensive way to move a product along a fixed path from one place to another. The forces of gravity and friction hold the package in place on the belt yet permit easy removal at destination. Also, the product may be held stationary while the belt keeps moving. This is useful to bank the product

Fig. 4.3 Automobile assembly line. (Courtesy General Motors Corporation, GM Assembly Division, Warren, Michigan.)

behind a gate, which provides for flexibility in the product flow since a unit can be released when the next station is ready or held when the station is occupied. Product banks of work-in-process inventory can temporarily support production flow when the standard times per station are unequal.

4.3.4 Overhead Conveyors

A conveyor over the work area has several advantages. The material can be hung on hooks or other devices. It makes use of the building's structure to carry heavy loads. An overhead conveyor is often used in mass-production painting. The product can be hung on hooks or moved automatically through the painting operation, either spray or dip. The paint can be applied in a separate room or chamber with no human present. This eliminates exposing the worker to toxic paint fumes. A typical installation is shown in Figure 4.4.

Fig. 4.4 Overhead conveyor for painting. (Courtesy Ransburg Corporation.)

A major use of the overhead conveyor is work-in-process temporary storage. The material can ride the overhead conveyor for a short time (up to a day in some companies) until it is needed for production or shipping. This avoids tying up labor and space for temporary storage on the shop floor.

Safety is an important consideration in overhead conveyors. OSHA regulations now require fence protection to keep material from falling on or hitting the worker.

Power-and-Free Conveyors

A power-and-free overhead conveyor enables the system to stop or divert items on an overhead conveyor. The carrier is temporarily detached from the moving conveyor and remains stationary until the carrier is reengaged. This allows for even product flow since product can be held until needed. It also provides for diversion of carriers to different destinations as desired. Various control systems are available so that conveyor operation does not require an operator's full-time attention.

4.3.5 Power Source

A power source is needed to move the conveyor and the material it carries. If the carrier is fixed, the material moves by gravity or by manual power. If the conveyor moves, power is suppled by electric motors. The material on the moving conveyor may be held in place by gravity or by some form of fixture or attachment.

An important consideration in many high-volume, mass-production industries is an alternate source of power in case the main power source is interrupted. This may be an auxiliary generator or other backup. The cost of downtime may be so great that a backup power source is economically justified. In some factories, material is moved manually or by fork-lift truck in order to keep production going when the moving line is shut down.

4.4 HOISTS, CRANES, MONORAILS

Hoists are mechanisms that lift product or equipment. They replace human labor by electrical power or by mechanically multiplying the human power applied to lifting.

A crane is "a machine for raising, shifting and lowering heavy weights by means of projecting swinging arm or with the hoisting apparatus supported on an overhead track" (1, p. 303). A monorail is a single rail track for a wheeled conveyance. It is usually mounted overhead with the vehicle suspended below the track.

4.4.1 Hoists

A hoist consists of a strong support (which may be part of the building structure); a hook or grasp; a chain, rope, or wire; a mechanism for raising or lowering the material; and a power source. A hoist may be operated manually by multiplying the human force applied or by electricity, hydraulics, or compressed air. The hoist may be mounted on an arm or a short, overhead track so the product can be shifted to another spot after lifting. A typical hoist is shown in Figure 4.5.

A digital readout device can be incorporated into a hoist as shown in Figure 4.6. The worker can read the weight of the load directly thus reducing the time required and increasing the accuracy.

4.4.2 Fixed Cranes and Derricks

Some cranes are fixed into position if the materials movement is repetitive and covers only a short distance. The major advantages of a fixed crane are a stronger mount and no time and expense moving it from place to place. It can be designed to handle heavy weights and is efficient and cheap in suitable applications.

A derrick is a framework over a drill hole (e.g., an oil well) for supporting the boring mechanism for hoisting. The term is sometimes used interchangeably with the term crane.

Fig. 4.5 A typical hoist. (Courtesy Ratcliff Company, Inc.)

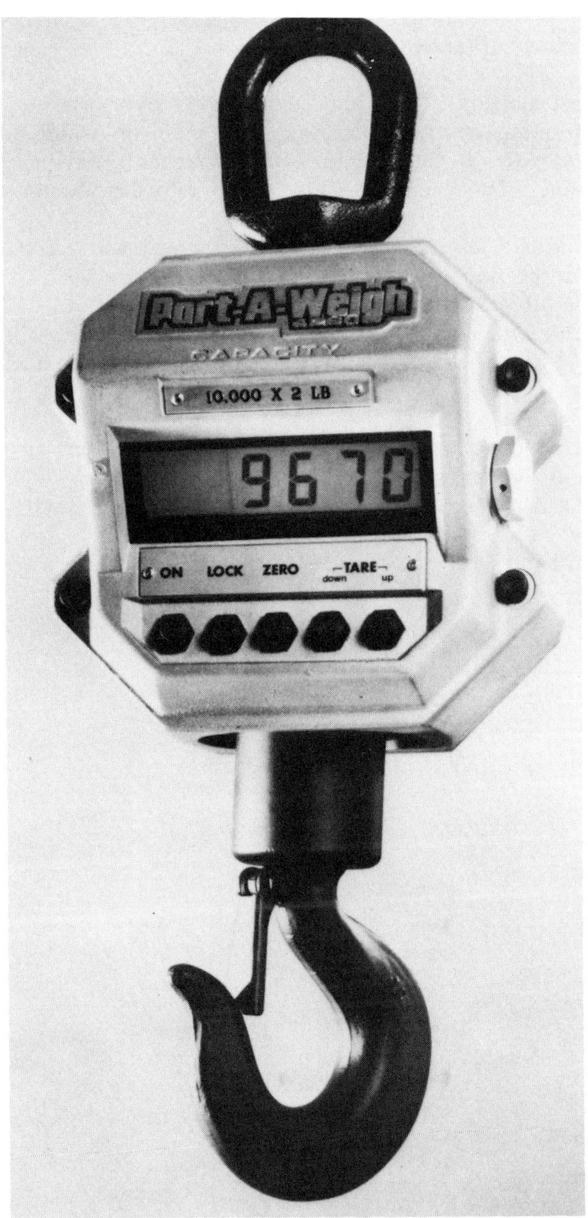

Fig. 4.6 Hoist with a digital weight readout. (Courtesy Measurement Systems International.)

4.4.3 Traveling Overhead Cranes

A practical and efficient method of moving heavy loads over the factory floor is the traveling overhead crane. The rails on which it moves are an integral part of the building structure and thus can support heavy loads. The crane can also move laterally along the traveling carrier.

Since it lifts the material above the top of floor equipment, it can move easily in a direct line to any spot within its range without detour. It does require adequate controls to operate and to avoid accidents and collisions.

4.4.4 Portable Cranes

A portable crane can be moved anywhere in the factory or yard as needed. It may be mounted on a truck or set on the floor or ground temporarily. It does have the disadvantage that care must be taken to ensure that it is stable for the loads carried. A typical truck-mounted crane is illustrated in Figure 4.7.

Fig. 4.7 Truck-mounted crane. (Courtesy RPC Corporation.)

4.4.5 Monorails

As the name implies, a monorail is an overhead track upon which the carriers or hoists move. The track is usually an I beam to provide strength and to minimize deflection. The monorail follows a fixed path, although provision can be made for sidings, switches, turntables, and other devices to divert carriers from the main track. A monorail is good for many processing operations such as meat packing. The animal is hung on the monorail carrier and moved from station to station as each operation is completed. Monorails can also be used for assembly and for transporting product within the plant. Another use is a rail extension outside the building over the dock to carry materials to and from trucks and rail cars.

4.4.6 Elevators

An elevator is a platform or caged hoist that moves material or personnel from one level to another. Modern practice is to avoid elevators in factories or warehouses if at all possible. An elevator is an expensive bottleneck and adds substantially to materials handling costs. In some older multistory buildings, an elevator is essential. However, its use should be minimized in designing the materials handling system.

An example of the high cost of elevators is shoe manufacturing. Industry engineers estimate that the cost of making a pair of shoes is 25–50¢ more in a multistory building than in a one-store structure. This is a fatal disadvantage in the highly competitive shoe industry. In many older cities, many multistory factories have been abandoned or converted to other uses. They are obsolete and too expensive to operate as factories.

4.4.7 Other

An example of a specialized materials handling equipment is the Hydra-Lift Karrier®, illustrated in Figure 4.8. The mechanism lifts a full 55-gallon drum and tilts it to the desired position. It can be easily maneuvered by the operator and is a flexible way to handle a drum requiring individual treatment and destination.

4.4.8 Power Sources

The power to operate hoists, cranes, and monorails can come from one of several sources.

1. Electricity
2. Manual
3. Pneumatic
4. Gravity

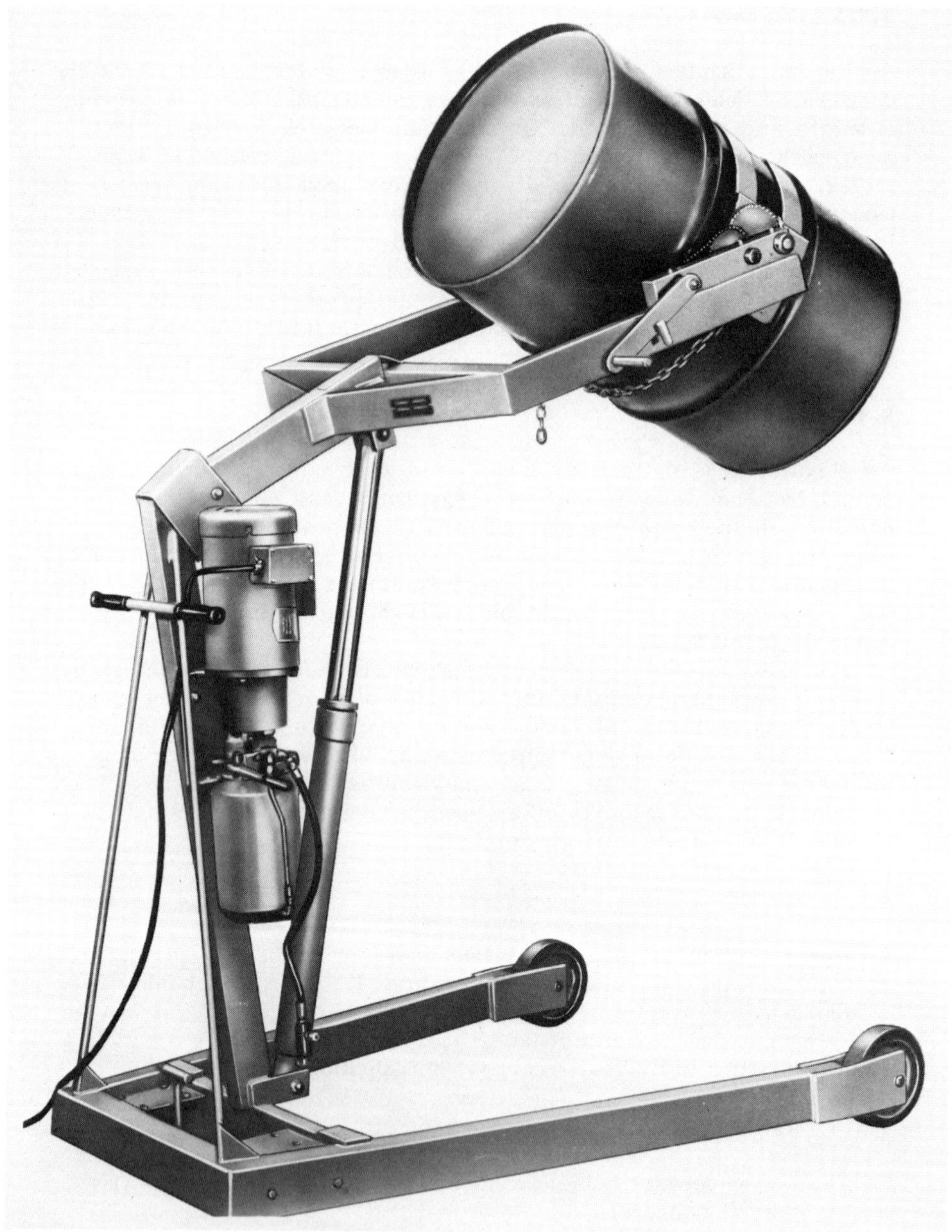

Fig. 4.8 Hydra-Lift Karrier®. (Courtesy Morse Manufacturing Co., Inc.)

Electricity

Electricity is by far the most common source of power for materials handling equipment. It is cheap, versatile, and flexible. It can be transmitted to any location where needed within the plant. Various types of motors are available for different applications.

Manual

Some simple hoists are operated by human power. This is primarily for lifting in applications not suited to electric power. The worker's force is multiplied through a system of ropes or wires and pulley. This method is slow but the investment is also small.

Pneumatic

Pneumatic power is useful for moving certain bulk materials in an enclosed pipe system. It eliminates the danger of sparking.

Gravity

Gravity is the least expensive power source if it can be used. Some bulk processing plants are built on a hillside so the product can start at the top level and move through the processing stages propelled only by gravity.

4.4.9 Controls

Hoists, cranes, and monorails need a control system to direct operations and to avoid accidents and collisions. These controls may be simple manual devices operated by the worker or elaborate electronic systems run automatically from a remote location. Each control system must direct the operation positively, quickly, and efficiently and must have provision for avoiding collisions and accidents.

Push Button

A crane or hoist may be controlled by the operator using a push-button box. The box is connected electrically to the operating mechanism, permitting quick and positive response. This type of control is useful for simple power hoists where the worker is close to the operation and can take action quickly. One is shown in Figure 4.9.

Radio

Radio is a convenient link between a worker on the ground and the operator of a large overhead crane, derrick, or hoist. Radio permits two-way communication for directions and cautions. The person on

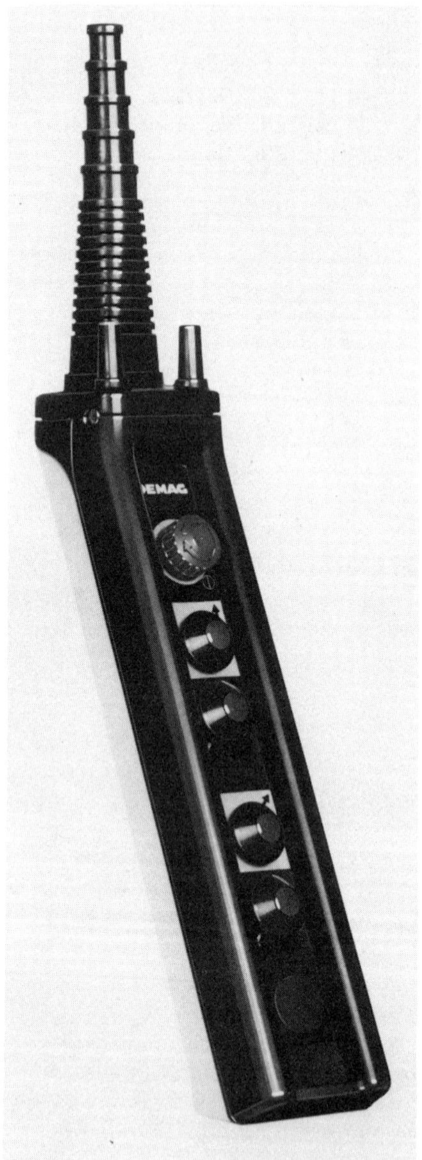

Fig. 4.9 Pendant control for hoist. (Courtesy Mannesmann Demag Corporation.)

the ground can signal completion of securing or releasing the load and warn of possible obstructions. The crame operator can query floor workers. The radio has the advantage of no wires, but must be free from excessive noise, static, and interference.

Computer

A recent development in materials handling has been the introduction of computers for controlling and recording the operation of a materials handling system. The computer system collects information on the status of the system and issues appropriate commands to the system elements. It can also be programmed for subroutines for part presence verification, collision avoidance, quality testing and recording, emergency signals such as shutdown, and other factors involved in the system operation. The investment required for a computer-operated materials handling system is high. This limits it to high-volume applications that justify the expenditure.

Manual

The worker can operate a simple mechanical hoist by manual controls. These are cheap, reliable, and relatively maintenance free. They depend on the experience, skill, and diligence of the operator.

Collision Avoidance

Material handling is a major source of industrial accidents. Every materials handling system should have safety designed into its operations. Safety precautions differ with the nature and complexity of the system.

However, two important elements of safe operation cannot be designed physically into a materials handling system. They are good supervision and personnel training. These are essential to safe operation of any factory and yet are often neglected.

4.5 INDUSTRIAL TRUCKS

An industrial vehicle is a self-powered, individually operated conveyance for moving materials or persons. Its flexibility, efficiency, and economy permit its effective use in many industrial applications. Industrial vehicles are readily available in standard models; a custom-built vehicle can be purchased for a special-purpose application.

4.5.1 Fork-Lift Trucks

The most common industrial vehicle is the fork-lift truck. It has its own power supply, a driver, and is adapted to lifting, transporting, and setting down a wide variety of materials and products.

Standard Models

A standard fork-lift truck is illustrated in Figure 4.10. Since it carries its own power plant, it is not limited to a fixed track or power supply. It is designed to hold and move loads safely and efficiently. Driving the fork-lift truck is similar to driving an automobile or small truck. Standard forks will fit standard pallets of various types. If the company can use the standard model, it is cheaper, has a shorter delivery schedule, and is easier to service and maintain.

High-Lift Trucks

Because of the increasing cost of land and building construction, high-rise storage is becoming more common in American industry. The cost per additional cubic foot is usually much less in the vertical direction than in the horizontal. Storage heights of 20 ft can be serviced easily.

The fork-lift truck is modified for high-rise storage. Extendable and retractable masts enable the operator to lift the load to the desired height and place it in the assigned rack or on top of the stack if no racks are used. The lift is also designed to maintain the center of gravity of the load toward the center of gravity of the vehicle to minimize the danger of tipping or spilling. A typical high-rise application is shown in Figure 4.11.

Side-Loading Trucks

High land and construction costs have led to another fork-lift truck development—the side-loader. This eliminates the need for turning room in the aisles. Only enough clearance for the truck itself and its load need be provided.

Special Attachments

Attachments are available for specialized fork truck applications. Prongs are used to transport sheet coils such as steel. Grippers are designed to grasp rolls of newsprint and cartons not mounted on pallets. A few types are shown in Figure 4.12.

Power

Three types of power are used for fork-lift trucks.

Fig. 4.10 Fork-lift truck. (Courtesy Eaton Corporation Yale Industrial Truck Division.)

1. Electricity. Battery power is efficient, nonpolluting, and quiet. The weight of the battery serves as a counterweight to the load. Its use is limited by high cost and the necessity to recharge the battery. Some countries require that all fork-lift trucks used indoors be battery powered.
2. Gasoline. This is the same type of internal-combustion engine that powers automobiles and trucks. It is economic, efficient, and does not need to be immobilized for battery recharging.
3. Liquefied petroleum (LP) gas. A vehicle can be powered by an internal combustion engine that runs on LP gas. This emits less pollution than does a gasoline-powered engine and is relatively efficient. A controlling factor in its use is the relative cost of gasoline and LP gas.

Safety and Training

Operation of a fork-lift truck is in many ways similar to the operation of an automobile or pickup truck. Also similar is the potential for

Fig. 4.11 High-rise fork-lift truck application. (Courtesy Eaton Corporation Yale Industrial Truck Division.)

accidents. In addition to conventional driver training, the operator must be instructed in safe handling loads, operating in confined factory spaces, and avoiding damage to personnel, equipment, or product.

4.5.2 Hand Trucks

For operation in limited areas with smaller, less frequent loads, hand-operated trucks may be optimal. The hand truck is pulled and operated manually. The operator inserts the prongs or platform under the load (or pallet), jacks the platform up off the floor, moves it to the destination, and lowers the load to the floor for release. Hand trucks are inexpensive, durable, versatile, and easily maneuvered. Similar trucks are available with self-contained power for larger, heavier loads.

Later developments include adding powered lifting and lowering of the forks, power for vehicle movement, and a place for the operator to ride. Necessary controls have been added. Power is furnished by electric batteries. An example is shown in Figure 4.13.

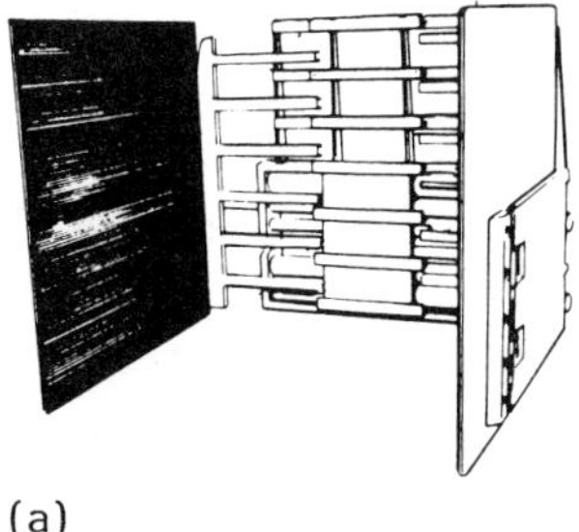

(a)

(b)

Fig. 4.12 Specialized attachments for a fork-lift truck (a) carton clamp (b) scoop (c) roll clamp (d) dumper/upender. (Courtesy Electroweld Company, Limited.)

4.5.3 Industrial Trains

Lifting and moving a single load with a fork-lift truck is versatile and quick. However, carrying a load on a vehicle is less efficient than pulling it. If volume warrants, an industrial tow tractor can pull a train of cargo vehicles at a low cost per unit. The cargo vehicles can be general or special purpose to suit system needs.

4.5.4 Special-Purpose Industrial Vehicles

Special-purpose vehicles are available either as individual vehicles or as part of a tractor train. These include:

1. Personnel carriers. These can move one or several employees quickly and safely to different locations throughout the plant.

(c)

(d)

Fig. 4.12 (Continued)

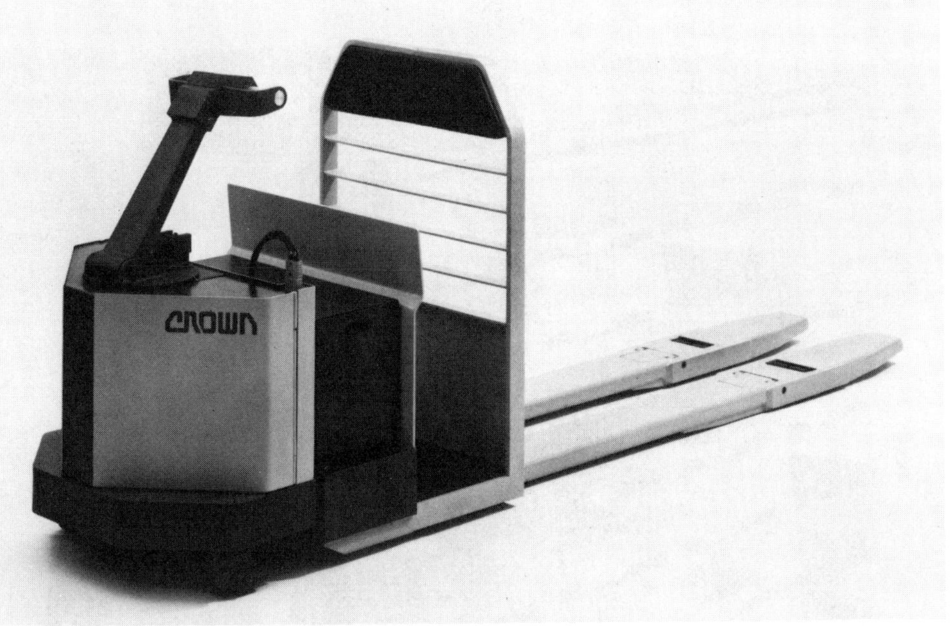

Fig. 4.13 Pallet truck with space for operator. (Courtesy Crown).

2. Maintenance vehicles. These are fitted with special equipment for various maintenance jobs. They can provide space for spare parts and supplies.
3. Janitorial vehicles. These carry materials and equipment for housekeeping services.
4. Agricultural vehicles. Many special types are available for use in agriculture or in agribusiness activities.
5. Other. If a company needs a unique vehicle and can justify the cost, there is usually someone who will make it to the buyer's needs and specifications. Figure 4.14 shows a Mailmobile®, a specialized delivery vehicle from Bell and Howell.

4.5.5 Automated Guided Vehicles

A large cost element of fork-lift truck operation is the driver's wages. In recent years, driverless industrial vehicles known as automatic guided vehicles (AGV) have been developed. These can operate in one of several different ways:

1. The vehicle may be hooked into a moving floor or overhead chain. If the chain runs in the floor, the operator places

Fig. 4.14 Mailmobile®, a specialized vehicle. (Courtesy Bell and Howell—Automated Systems Division.)

the vehicle over the moving chain, drops the engagement pin, and the vehicle moves off. To remove the vehicle from the chain, the operator lifts the pin.

2. The vehicle can follow electronically a strip of special paint on the floor or a wire embedded in the floor. Electronic impulses guide and control the vehicle. This technology has been improved and its use expanded in recent years. It is discussed in more detail in Chapter 17, "Automated Guided Vehicle Systems."

In all driverless systems, the vehicle must be fitted with collision-avoidance capability. Upon contact, the vehicle will come to a dead stop before any damage is done to personnel, materials, or structures.

4.6 POSITIONING, INSPECTING, AND WEIGHING EQUIPMENT

A fundamental principle of methods improvement is to combine operations whenever possible. Applying this to materials handling, the system should perform related operations as part of the materials handling function. Examples are positioning parts for the next operation, weighing and counting, and automated inspection.

4.6.1 Positioner

A positioner is a device that orients and positions the part for the next operation. If positioning can be done while the product is being moved, hand labor can be eliminated and costs reduced. A simple guide bar may be sufficient to turn the part or to steer it to the proper location.

Manipulators

A manipulator is a device that grasps an item and moves it to a new position or orientation. It may be a simple pincer or complicated robot. The purpose is to place the item in the right position for the next operation. Performing this function automatically eliminates hand labor and may reduce misplacement and other errors. However, the manipulator must usually be supplemented by controls that check correctness of the orientation and verify part presence or absence. A worker can do these functions easily; a machine must be carefully programmed to perform them.

A manipulator must usually be custom designed for the specific application. Cost precludes its use except on high-volume, repetitive operations. An example of a manipulator appears in Figure 4.15. Research and development of flexible robots is going on. When

Fig. 4.15 Manipulator. (Courtesy Sauder Energy Systems, Inc.)

developed, these robots can quickly and easily be programmed to accommodate different parts.

Upenders and Dumpers

Upenders and dumpers are specialized machines to unload materials handling carriers at a desired location. They are used for unloading bulk materials from a conveyor to another conveyor, a transportation vehicle, or a storage location.

Positioning and Transfer Mechanisms

In a fully automated production system, the product is moved automatically from station to station. A common feature is the transfer mechanism, which removes the item from one station, moves it to the next station, and positions it for the next operation. This requires precision in both pickup and placement. In some automotive applications, the placement must be within a thousandth of an inch or less for machining operations with a tight tolerance.

A transfer mechanism is normally custom designed for a specific application. This may require a year's lead time for tool design and acquisition. As a general rule, automatic transfer mechanisms are cheaper for high-volume repetitive manufacturing than are robots and other flexible mechanisms.

4.6.2 Inspection

Manual inspection has seldom been completely satisfactory. The work is repetitious and boring; it is difficult for a worker to maintain good production and quality. Several studies have demonstrated that 100% accuracy cannot be reached with manual inspection. If automatic inspection and quality assurance can be integrated into the materials handling system, results can be improved and costs reduced.

Automated Inspection

In many systems, inspection of parts, subassemblies, and final product has been automated. The inspection system measures each designated characteristic, compares it to the standard, indicates acceptance or rejection, and records the results. A number of measurements can be made simultaneously with good precision. This not only saves time and labor cost, it reduces the errors to which the human inspector is subject.

Integrating the Inspection Station into the Materials Handling System Flow

Much time and expense of traditional inspection is involved in manual handling of the product and inspection equipment. This is eliminated by automating the inspection operation. However, the inspection must be integrated into the entire material handling system to ensure even, economic product flow and to segregate the rejected from the accepted items. This requires the system designer to provide for:

1. Automatic flow and handling into and out of the inspection station.
2. Separating rejected items from the regular system flow. In some cases, the system may identify the rejects for later manual removal.
3. Verifying part presence, absence, duplication, and orientation.
4. Connecting the inspection station with the computer system which records and analyses the results and transmits necessary reports to those who need the information.

4.6.3 Weighing Equipment

Weight of an object or package is a needed measurement for many items. If weighing can be done automatically or combined with another operation, time and money can be saved. An example is a mailing machine which automatically weighs the package, computes the required postage, and prints the correct metered amount on the package. Modern equipment can weigh with excellent precision.

Weight Counting

A common technique is weight-counting small items. One or a few items are weighed separately from the rest of the lot. From this and the weight separately of the entire lot, the computer calculates the number of pieces in the lot. Although this may occasionally result in a small error, the loss is insignificant compared to the savings over manual weighing and counting.

Integrating the Weighing Operation into the Materials Handling System Flow

Again, integrating the operation into the materials handling system is essential to economic efficient operation. The system must:

1. Move the materials to be weighed onto the scales
2. Perform the weighing quickly and accurately
3. Record and transmit the data as desired
4. Move the material out of the weigh station to the next operation or to storage

4.7 OVER-THE-ROAD MOTOR VEHICLES

Motor vehicles are often part of an industrial materials handling system. Standard vehicles are preferable to specially designed ones since they are cheaper and more readily available. However, special bodies and equipment can be obtained for a price which may be far less than the saving obtainable. An example is a simple unloading crane mounted to a truck for unloading material at a building site. Since most readers are familiar with motor vehicles of all types, no detailed descriptions will be given here.

4.8 RAILROAD EQUIPMENT

Some heavy industries such as steel use railroads as part of the materials handling system. This is appropriate for industries that use large quantities of heavy bulk material shipped by rail. Bringing the cars into the factory to the point of use saves the cost, trouble, and possible damage from rehandling and transhipment. Some companies own and operate the railroad within the plant; others use the equipment and services of the railroad company.

Specialized railroad equipment can be obtained and is available to large shippers. One example is the specialized rail car for hauling automotive subassemblies and completed vehicles. Rail cars may be fitted with other materials handling equipment such as cranes, hoists, and scoops.

4.9 MARINE CARRIERS

Transportation by water is an ancient way of moving large volumes of material. Prior to the invention of the railroad, overland transport was difficult, slow, and very expensive. Water was slow but materials could be moved long distances at low cost.

Water transportation is still inexpensive and often is the most satisfactory method for moving large volumes of bulk raw materials and for moving goods to locations that are impossible to reach by land. Most large cities of the world are on navigable waterways.

Two major categories of cargo are moved by water—bulk and container. Bulk product includes coal, oil, cement, and timber. These are handled by specialized equipment and vessels. Manufactured goods normally move by containers. These are steel boxes of various standard sizes and specifications. The dimensions of the most common size are close to those of the standard over-the-road cargo trailer body. Containers can be handled quickly and easily in ports that have the specialized equipment and can move overland on rail or flat-bed trailers.

4.9.1 Ocean-Going Vessels

Vessels for ocean shipping must be designed to withstand the rigors of storms, waves, high winds, and other hazards of the open ocean. Some are specialized, such as large tankers for carrying crude oil from producing areas to refineries and consuming areas. Some are designed to handle ore, grain, or coal. Others are general cargo for container shipment.

4.9.2 Seagoing Containers

A seagoing container is very similar to the cargo section of an over-the-road trailer in construction and dimensions. Many are moved overland by trucks using a special frame with wheels as a trailer to carry the container. The dimensions and specifications have been standardized internationally. One is illustrated in Figure 4.16. Figure 4.17 shows a seagoing container being carried by a straddle carrier. Figure 4.18 depicts a straddle carrier moving a large pleasure cruiser.

Fig. 4.16 Seagoing container. (Courtesy Fruehauf Corporation.)

Fig. 4.17 Seagoing container being moved by a straddle carrier. (Courtesy RFC Corporation.)

Ideally, the container is packed and sealed by the originating shipper, moved to the port, loaded on ship which sails to the destination, unloaded intact, transported to the consignee, and unloaded. If the container can be sealed from origination to destination, handling time, damage, and pilferage are greatly reduced. Most ports have the specialized materials handling equipment to handle containers quickly, cheaply, and safely.

4.9.3 Lake Vessels

Ships that operate exclusively on inland lakes are built to withstand the forces prevailing where they operate. Ore vessels operating on the Great Lakes must survive Lake Superior storms, which many sailors regard as more vicious than ocean storms. On the other hand, vessels for milder, shallower lakes are built with less draft

Fig. 4.18 Straddle carrier moving a large pleasure cruiser. (Courtesy ACME Hoist Incorporated.)

and protection against storms. Inland lake transportation is high volume in some areas such as the Great Lakes of North America.

4.9.4 River and Canal Boats

Traditionally, entire nations and trade regions developed along navigable waterways. Egypt developed along the Nile River. Rivers have the advantage of easy water transport and lack of high waves and heavy storms. On the other hand, river transport can be limited by falls, rapids, shoals, sandbars, seasonal rise and fall, and shallow water.

Vessels for operating on rivers and canals are built with shallow draft and low freeboard when loaded. They are slow since high speeds on rivers and canals are impractical. However, they can carry large bulk and heavy loads cheaply. One of the busiest rivers in the world is the Monongahela with its coal and steel traffic.

4.9.5 Barges and Lighters

Barges and lighters are shallow-draft, boxlike vessels used for cargo transport in protected waters such as bays, rivers, and canals. They are not self-propelled, so they must be moved by a tug with small, powerful engines. On most rivers, one tug will move several barges fastened together. This is slow but cheap.

4.9.6 Small Craft

Smaller craft can be used for specialized tasks or for operation in restricted waters where larger vessels cannot go. An example is the lighter craft used for ship-to-shore transfer in shallow harbors. Many specialized models are available. Custom vessels for special situations can be built.

4.9.7 Mobile Floating Equipment

For some applications, materials handling equipment such as a large crane may be mounted on a barge for moving around a bay or harbor area.

4.10 AIRCRAFT

Aircraft can be part of the materials handling system. One garment company has a daily air round trip from the U.S. mainland to Puerto Rico. The outgoing cargo is cut and partly finished garments. The garments are completed in Puerto Rico and returned by air to the mainland.

A limiting factor on air transportation is the time and expense of ground transportation to and from the airport. Often, air time is such a small fraction of the total trip that savings from faster airplane speed is more than offset by the time and expense of the ground transport.

Since the reader will be familiar with the types of aircraft available, no detailed description will be given here.

4.11 UNIT LOADS AND UNITIZING EQUIPMENT

A unit load is a standardized combination of smaller items into a larger integrated group which can be handled as a single item. Figure 4.19 illustrates a typical unit load. The high cost of manual labor has made individually handling small packages and items prohibitively expensive. If a number of items can be handled as a unit, materials handling costs are reduced by moving larger loads; eliminating unloading and reloading; cutting travel time; using space more efficiently; facilitating shipping, transport, and receiving; and reducing inventories. Raw materials, work-in-process, and finished goods can be unitized.

Fig. 4.19 Typical unit pallet load.

The unit load is built on a standard platform or container. It can be designed for a single large-volume product or for a wide variety of products. The system may use a simple pallet on which the product is stacked and tied down or an open-top container into which product can be placed in any orientation. Some single-purpose standard containers are provided for liquids, gases, or toxic materials. An example is the standard cylinder for welding and other gases.

Materials handling is cheaper, faster, and more efficient if automated equipment can be used. However, this works best if each item transported is identical. A unit load is used for small and medium-sized products. A further development is equipment that receives items individually off the conveyor line and arranges them into the desired unit load. Since this is done automatically, most manual labor is eliminated.

Production control is facilitated by standard unit loads. It is easier to count and keep track of work-in-process by number of loads than by number of product if each load consists of the same number of items.

4.11.1 Unit Load Design

Unit load design should meet the following criteria:

Size and Weight

The unit load size must be such that it can be handled easily and economically with modern equipment. Usually, this means a 48 in. by 40 in. pallet with a product height of about 4 ft. Some unit loads may be bigger or smaller to meet other criteria and the characteristics of the product.

The weight of the unit load must be kept within the capacities of the materials handling equipment and the storage facilities. Unit loads of some high-density materials, such as steel, flour, stone, and brick, are smaller than optimum size in order to keep the unit load weight within limits.

Compatibility

The size, shape, and weight of the unit must be compatible with the equipment and facilities available. It is usually uneconomical and inefficient to specify different size and shape unit loads for different items within the same materials handling system.

Damage Control

The unit load must be designed to minimize damage to the product and equipment. Errant fork lift prongs are a frequent source of damage to product; heavy steel bins will minimize this danger. Damage from crushing must be avoided if the unit loads are to be stacked in high columns.

Arrangement

The individual items must be arranged to minimize the chance of spilling and to facilitate the use of automatic palletizers. Usually layers of packages are arranged so that each package spans more than one package on the next lower level. This helps hold the load together. See Figure 4.20.

4.11.2 Containers for the Unit Load

The unit load calls for a standard base or container. Among the possibilities are:

1. Pallets
2. Skids
3. Slip sheets
4. Containers
5. Self-contained carton

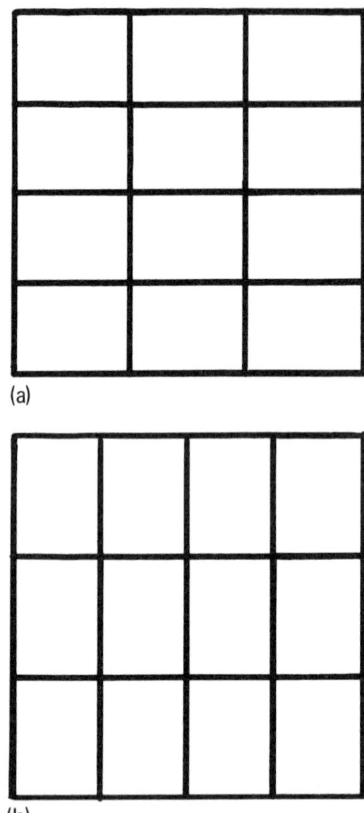

Fig. 4.20 Arrangement of unit pallet load (a) odd-numbered layers (b) even-numbered layers.

Pallets

A pallet is a standard platform on which material is placed for storage and movement. These are platforms with an upper and lower flat surface with space between for the forks of an industrial lift truck. The most popular material is oak because it furnishes satisfactory strength and service at low cost. Other materials are plastic, steel, aluminum, and corrugated board. Most pallets serve several years before they must be discarded. Figure 4.21 shows a typical pallet.

Skids

These are similar to pallets except that they have no bottom surface. They are economical and rugged but are not satisfactory for stacking one upon another. A common form is made from a single piece of steel with the runners bent on a pressbreak.

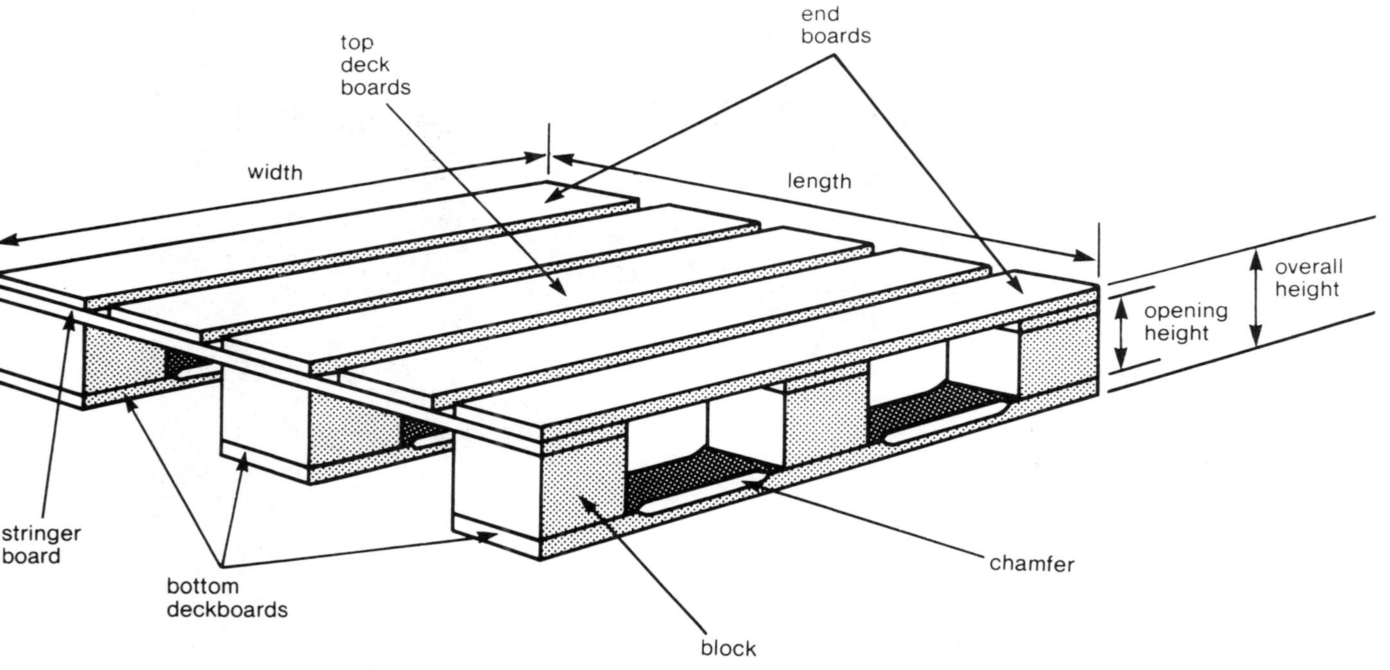

Fig. 4.21 Typical wooden pallet. (Courtesy National Wooden Pallet and Container Association.)

Slip Sheets

A slip sheet is a single sheet of heavy, strong corrugated fiber board or plastic. The unit load is pulled onto the slip sheet and the forks of the lift truck slide underneath. Slip sheets reduce the cubage under that required for a palletized load. A special truck for handling slip sheet unit loads is shown in Figure 4.22.

Fig. 4.22 Push-pull truck for slip sheet loads. (Courtesy Long Reach Manufacturing.)

Containers

A standard container may be used for unit loads. A bin of heavy steel mounted in a pallet will hold such items as loose, irregular metal castings. An example is in Figure 4.23.

Many varieties of open-top containers are available for materials handling storage. The weight of the parts determines the size. Tote boxes for steel parts are smaller than containers for clothing, shoes, and other soft goods. Containers can be designed for stacking or for combining into larger unit loads.

Some containers are specialized for a single product. Examples are compressed gas cylinders and paint and oil drums. Special racks can be built for mass-produced parts and products. Traditionally, a shoe factory uses a wooden rack that holds 18 pairs of shoes.

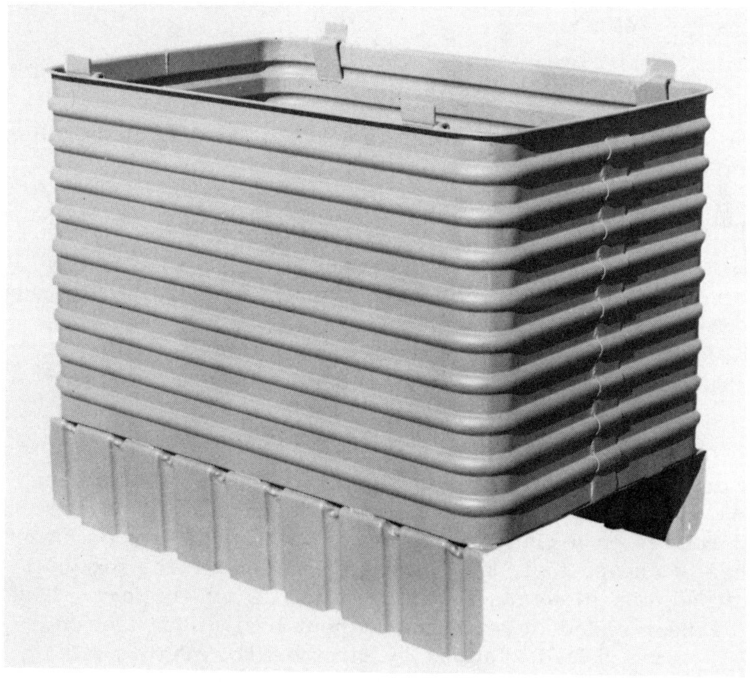

Fig. 4.23 Skid with container. (Courtesy Streator Dependable Manufacturing Company.)

Self-contained Carton

A unit load may have no base or container. An example is a large appliance such as a stove or washing machine. Given a carton of sufficient strength and proper design, a separate base outside the carton is unnecessary. The carton may be handled with special attachments to a fork-lift truck.

4.11.3 Palletizing and Unitizing Machinery

Palletizing or unitizing equipment is available to combine parts or packages into uniform unit loads. These remove product from the end of the production line and stack it in unit load arrangement automatically with no manual handling. The machines are particularly useful for product packed in standard corrugated cartons.

A palletizer is commonly used with high-volume production of uniform packages such as cartons of a food product. The palletizer is integrated into the production and packaging line so that the palletizing is done without manual labor. The system may call for a fork lift to remove loads. In some instances the unit load may move by conveyor to the ultimate destination within the plant. Figure 4.24 shows a palletizer.

4.11.4 Securing the Unit Load

There are several methods for securing the unit load against spilling, items falling off, and other hazards. These include:

Strapping

A flexible steel strap wrapped around the load and joined with a special fastener will hold the unit load securely in place. Care must be taken that the strap does not bite into the load and cause damage. Machines to strap automatically are available. Other materials such as gummed-paper tape and fiber tape can be used.

Film and Stretch Wrapping

A recent development is wrapping the unit load in a transparent plastic film. An automatic machine wraps the plastic around the load as it rotates on a platform. A sealer fastens the plastic cover. The method is cheap, fast, and less likely to damage the product. The plastic adjusts to contours and irregularities of the load. The film can be heat-sealed or stretched without heat under tension. Figure 4.25 shows a load wrapped by stretch film. Figure 4.26 pictures a unit load of irregular objects approaching the wrapping mechanism.

Fig. 4.24 Palletizer. (Courtesy Columbia Machine, Inc.)

Adhesives

Special adhesives are available for holding a pallet load together during handling and shipment. These adhesives are strong in compression and weak in tension. They will hold the load together yet the individual packages can be separated easily for distribution. This is not suitable for rugged duty but can be an inexpensive method for securing a load of lighter items not subject to rough treatment.

4.11.5 Integrating the Unit Load into the Materials Handling System

The materials handling engineer must integrate the unit load into other facets of the production and material handling system. The package must be designed so that it can be handled automatically, will be compatible with the unit load, and will protect the contents from damage, loss, or deterioration.

Fig. 4.25 Enclosing a unit load with stretch wrap film. (Courtesy Aktron Corporation.)

Fig. 4.26 Pallet wrapping machine using stretch plastic. (Courtesy Herbert S. Warmflash and Associates, Inc.)

The current rate structures of common carriers such as railroads and trucks require or strongly favor carload or truckload lots. The manufacturer loads the car or trailer with his unit loads, which move without transfer to another vehicle to the destination, where the unit loads are removed from the vehicle by the receiving organization.

4.12 STORAGE EQUIPMENT

The ideal storage is none at all. As usual, the ideal is seldom attainable in practice. Even so, inventory and storage should be kept at a minimum.

Every materials handling system must include some provision for storing material. In a fully automated production system, only storage for raw materials and for finished product need be provided. All work-in-process is in motion through the manufacturing system from start to finish. In some manufacturing, banks of partly finished work-in-process may be necessary. The system must provide space and equipment for this storage.

An industrial engineering group was asked by a supermarket chain to determine the optimum size of the storeroom in the back of each store unit. Their recommendation was "Have no storeroom at all and remove those already installed." The investigation showed that the store manager filled the storeroom no matter how big or small. If the storeroom was abolished, incoming goods went directly to the shelves where they were available for sale to customers. Otherwise, goods would be put into the storeroom, where they contributed nothing to current sales. Unfortunately, complete abolition proved impractical; space was needed for empties awaiting return, damaged goods awaiting disposal, and similar items. The chain did reduce the size of the storeroom and prohibited putting salable goods in the storeroom. The axiom is "If storage space is provided, someone will fill it." Storage space should be kept at the minimum necessary for efficient operations.

At a materials handling workshop, the representative from a large and successful manufacturer reported that a major problem was mislocation of or mislaying work-in-process inventory on the shop floor. The material was there, but was not available for use because it could not be located. The storage and materials handling system should be designed so that location of all work-in-process material is known at all times. Often this can be accomplished by careful design of automated handling and storage, which minimizes the chance of losing materials on the shop floor.

A key factor in efficient operations is high utilization of available floor space used for storage. Each square foot and cubic foot of storage costs money; the layout and operations should be designed to maximize utilization of the space.

One method of increasing utilization is high stacking. It costs much less per cubic foot to build up than to expand laterally. If the product is light enough to permit high stacking, considerable floor space can be saved by building racks of several tiers or by stacking unit loads higher. Fork lifts with high reach capacity are available to lift and bring down unit loads from heights up to 20 ft. Figure 4.27 illustrates a typical pallet rack installation.

4.12.1 Cantilever Racks

For many types of material, such as metal rods and bars, cantilever racks are efficient and economical. They are supported only on one end, leaving the other end open for placing and removing the long bars. In combination with side-loading trucks, they use space efficiently. Storage can be outdoors if the material and climate permit. An example is shown in Figure 4.28.

Fig. 4.27 Pallet rack. (Courtesy Equipto.)

Fig. 4.28 Cantilever storage rack. (Courtesy Cal-Rak Corporation.)

4.12.2 Flow Racks

A flow rack is designed so the material can be inserted at one side and will move by gravity to the other where order picking takes place. It is particularly suited to handling high volumes of a small number of fast-moving items. A sketch of a typical installation appears in Figure 4.29.

4.12.3 Shelves, Bins, Drawers, Stacks

Many organizations store slow-moving items such as tools and supplies on shelves, in bins, drawers, or stacks. They use space efficiently in comparison with floor storage. Major problems are holding obsolete items after their usefulness has ended, keeping track of location, and sometimes inefficient use of storage space.

Shelves are cheap and easy to install. The stored items are visually accessible. This makes locating the item and taking inventory easier. Small containers such as cartons and boxes can be stored on shelves. Identifying labels are visible from aisles. On the other hand, space may be inefficiently used if a small item is

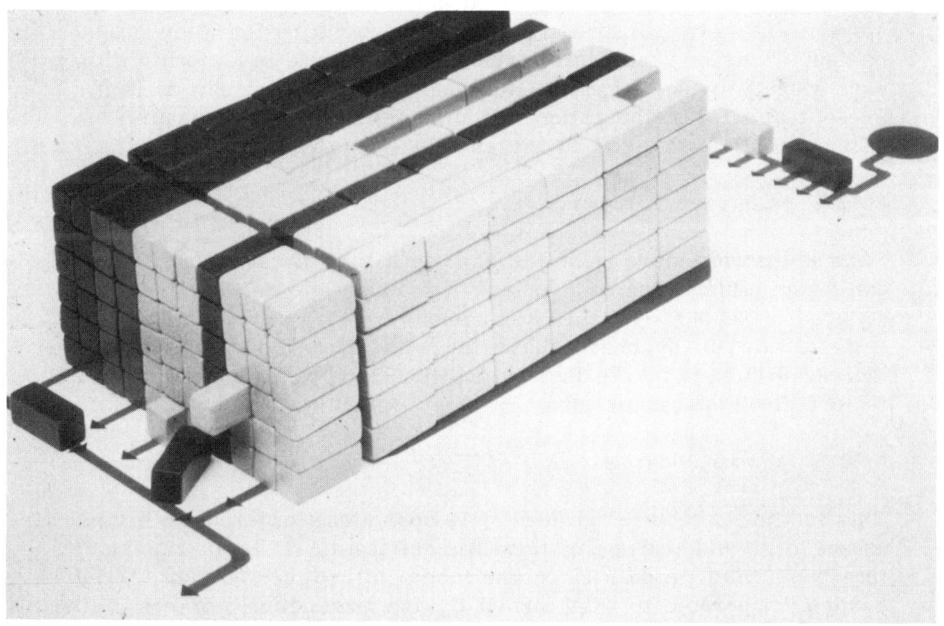

Fig. 4.29 Flow storage installation. (Courtesy Interoll Corporation.)

stored on a shelf capable of holding a much bigger item. Flexible shelving permits moving shelves up or down to accommodate the size of the items.

Bins are used for storing small items which could spill if placed on an open shelf. Nails, machine screws, nuts, office supplies are often placed in bins. Drawer storage is used when the items are subject to pilferage or must be protected from dust and other environmental damage. Drawers must be properly labeled to facilitate access and to avoid "losing" material because its location is unknown.

Some items can be simply stacked on the floor or ground. This is easy. Many large items suffer no damage when stacked. Care must be taken that the stack is stable and that the floor load limits are not exceeded.

4.12.4 Carousels

A carousel is a rotating or circulating storage device. The worker stays in one place while the needed item comes to the work station. A common application is the circulating conveyor used in dry-cleaning establishments to bring the customer's finished work to the counter rather than having the clerk walk through a set of racks searching for the desired garment.

Carousel storage is cheaper to install and operate than automatic storage and retrieval systems or miniload systems for many companies. This is particularly true for low-volume inventories with a wide variety of small items. Because the number of carousels in present and future operation is increasing, a separate chapter (Chapter 16) is devoted to the subject.

4.12.5 Conveyor Storage

Some industries store materials and work-in-process on a moving conveyor. This eliminates manual unloading and reloading yet the material is readily available when needed. The conveyor may be designed to run overhead above the work area and equipment. The increased cubage in the factory may be cheaper than expanding the factory floor space laterally.

4.12.6 Order Picking

This is the function of gathering various items ordered by a customer, assembling, and packaging them for shipment. It is an expensive, time-consuming process since the labor, although unskilled, receives a wage comparable to that earned by the production workers in the same plant. If the company must perform this function, the layout must be designed to accommodate the work.

Special vehicles and equipment have been developed to facilitate the order-picking process and improve its efficiency. Figure 4.30 shows a man-aboard vehicle for order picking from multilevel stack warehouses.

Many companies reduce order picking by selling only to distributors, who will handle the order picking, or by selling in minimum quantities which reduce the individual order picking.

4.13 CARLOAD AND TRUCKLOAD LOTS

Most transportation companies give substantial rate reductions in dollars per hundred pounds to a shipper who purchases an entire railroad car or truck trailer for transportation between two points. This may save the company considerable money over shipping less than truckload lots. About 30 years ago, General Electric consolidated all its large appliance manufacturing in Louisville, Kentucky. A

Fig. 4.30 High-rise fork-lift truck with platform for order picker. (Courtesy Drexel Industries.)

primary advantage was the increased ability of the company to ship carload lots by combining different appliances destined to the same distributor into one car or truckload lot. Also, the location was near the center of population of the continental United States; this tended to keep overall shipping costs to a minimum.

Full-load lots have other advantages, such as faster service, fewer terminal delays, decreased exposure to pilferage and damage. Frequently, the item can be purchased at a lower unit cost because of the supplier's savings in labor in making up and shipping a full-load lot.

The railroads have gone even further with high-volume shippers. A company can buy part or all of a unit train for bulk transportation between two points. An example is shipping large quantities of coal from mine to an electric power generating plant.

REFERENCE

1. By permission. From Webster's Ninth New Collegiate Dictionary © 1986 by Merriam-Webster, Inc. publisher of the Merriam-Webster® Dictionaries.

5

Criteria for Designing a Materials Handling System

5.1 INTRODUCTION

Before we start designing a materials handling system, we need to establish our objectives, what we are trying to accomplish, and our criteria, the measures by which we judge the extent to which our proposals meet our objectives.

5.1.1 What Is a Criterion?

The dictionary defines a criterion as "a standard on which a judgment or decision may be based" (1, p. 307). It should be quantitative if possible. If not, it should be a clearly defined parameter usable for objective comparison of alternative materials handling systems. An example is cost savings. The annual cost savings over the present method is determined for each alternative. A comparison of these cost savings shows which proposed system is superior by this criterion.

5.1.2 Why Defining Your Materials Handling System Criteria Is Important

Without clearly defined objectives and criteria, the designer has no basis for choosing among the many primary and subordinate alternatives available for his system.

The designer uses his criteria to evaluate alternatives in system and subsystem design. First, he must choose among the general types available. For example, should it be based on fork-lift trucks or conveyers? The designer bases his decision on the extent to which the fork lift and the conveyor systems meet the criteria for the system. If the fork lift is chosen, he must then compare various types of fork lifts available. Should fork lifts be standard or high-lift? Have special attachments? After the type of fork-lift truck is determined, which make and model is best? All these decisions are based on comparing alternatives for their ability to meet the system criteria.

5.1.3 Relationship of the Criteria to Objectives

The design process will operate best when each criterion is related to one of the system objectives. An example is the system operating capacity that relates to the objective of moving production effectively. Another is equivalent uniform annual cost, which is a measure of achieving lowest cost.

5.2 CRITERIA MEASUREMENT

A criterion starts as a statement of the parameter for comparing alternatives. For example, one might say "Calculate the capital investment required." To be useful, this statement must be converted to a mode of measurement. This criterion is easy. "Capital investment will be measured in dollars." A good criterion is one that can be expressed in quantitative terms such as dollars per unit, number of unit loads per hour, or percent uptime. Not all criteria can be expressed in numerical terms; some may require some surrogate calculations. Others can be expressed only in qualitative terms. Occasionally, a criterion that would be desirable cannot be used because there is no usable measure for the criterion.

5.2.1 Quantitative

The most useful criteria are those which can be measured in quantitative terms directly related to the system and its objectives. Examples are flow rates in unit loads per hour and operating cost in dollars per unit or per time period.

5.2.2 Scaling

Some important characteristics are not definable directly in quantitative terms. However, a scale can be used to quantify evaluation comparisons of a specific characteristic. A scale is set up with 100 equal divisions, with 100 being the best possible and 0 the worst. For example, flexibility for future changes might be scaled with the score for each alternative assigned according to the degree of flexibility as detemined by the analyst.

5.2.3 Zero-One

For some important factors, the proposed materials handling system either does meet the criterion or it doesn't. There are no intermediate values, no degrees of attainment. These are zero-one variables.

An example is conformance to zoning and building codes. Either the system meets these regulations or its doesn't. If it conforms, the system meets the criterion with a value of one. If not, the value is zero. It may be possible to secure a variance or exception; unless this can be obtained, the alternative is eliminated.

Management may impose a zero-one criterion. A typical example is "If the system won't pay back its cost in 3 years, forget it." Management can thus set limits on key variables as a preliminary sort mechanism. After the proposal has passed the sorting barrier, it may be evaluated on a quantitative basis for the final decision even though a zero-one barrier was imposed earlier.

5.2.4 Intangible

Some criteria are evaluated on a qualitative, not quantitative, basis. It may not be feasible to establish a meaningful quantitative measure for a criterion, or the cost of doing so may exceed its value in the analysis. These characteristics are called intangibles or irreducibles.

An example would be community relations. Although it might be feasible to establish an index for community relations, the cost would be high, the reliability and accuracy questionable, and interpretation by management, supervisors, and staff difficult.

5.3 COSTS

The most important criterion in most materials handling systems decisions is money. However, there are several different aspects to the money factor; each may call for a different approach.

5.3.1 Capital Investment

The amount of capital investment money needed to acquire the materials handling system is an important managerial consideration, especially in economically depressed times. Almost every enterprise has many more desirable investment opportunities than it has funds to invest. The fund limitation may lead management to favor a proposal that requires a smaller investment even though another would be more beneficial to the firm in the long run.

5.3.2 Annual Operating Expense

Another important financial consideration is annual operating expense. This is particularly important when the capital investment is low in relation to the total operating cost. Current expenditures such as direct labor and materials, power, and disposable containers are quickly reflected in the current accounts, unit costs, and bids to customers.

5.3.3 Combined Cost Techniques

Seldom can capital investment costs and annual operating expenses be considered independently. However, there are conceptual and interpretational difficulties in combining them into a single useful financial measure. In spite of these problems, most companies use an analytical method which combines all costs associated with an alternative into a single measure.

Present Worth

In the present-worth method, all expenditures associated with an alternative are converted to a single equivalent present amount. Investment costs are seldom a problem; they are already valued at the present worth. Annual expenses are discounted to the present equivalent amount by multiplying them by discount factors which compensate for the time value of money.

Present worth has two major difficulties:

1. The present amount is often misinterpreted as the amount needed for front-end investment.
2. It is not useful for comparing alternatives with different service lines.

Equivalent Uniform Annual Cost

In this method, all costs are converted to equivalent annual costs and then summarized. It can be used for comparing alternatives with unequal service lines. It is easier to interpret and explain than present-worth totals, since most managers are used to thinking in annual income and expense terms.

Rate of Return

The rate of return on investment is a useful method of comparing similar and dissimilar investments for profitability to the company. The net return to the company in increased revenues or decreased costs is expressed as a percent return on the investment required. It is used in organizational capital budgeting in which management is selecting a limited number of projects from a large number of varied investment proposals to maximize the return from the available funds for capital investment. These proposals can be for different types of capital expenditures.

Payback Period

A simple, widely used financial yardstick is the length of time required to recover the initial investment through net earnings or savings generated. It has the advantages of:

1. Measuring liquidity by the time required to return the original investment to the cash account.
2. It is simple to calculate and easy to explain.
3. It tends to be conservative in judging long-term commitments.

5.4 CUSTOMER SERVICE

A company's ability to serve its customers with on-time delivery, repair and maintenance services, minimum damage, and prompt attention to special orders and complaints is a key component of prosperity and survival in today's highly competitive economy. Better customer service may be the crucial factor in a decision to acquire an improved materials handling system. How do we measure customer service?

5.4.1 Time to Shipment

The time required from receipt of the order to shipment is a good measure of customer service. Present data are usually obtainable from existing sources; sampling may be used for speed and economy. The time for the proposed system may be estimated from parameters of the proposed system design. It may be desirable to use the percent of on-time shipments if there is slack time between the date the order is received and the required delivery date.

A large buyer can and often does insist that every shipment be on time. If so, the system must provide this service. This requirement calls for a high degree of reliability and a backup in case of system component failure.

5.4.2 Stockouts

In industries that ship from inventory, the number of stockouts may be a good measure of customer service. This is computed on the basis of the number of times an item ordered by a customer is out of stock, not the number of times an item reaches zero inventory.

5.4.3 Damage

Damage to product is a detriment to good customer relations as well as an expense. A good materials handling system will move goods efficiently with minimal damage. This measure is seldom obtainable directly from existing records since most damage charges are entered in an account that covers all damages or are buried in general overhead or in other unidentifiable accounts.

5.5 WORK-IN-PROCESS

Work-in-process inventory is a major investment for most companies. It is defined as product material which has been removed from the raw material inventory and put into production but which has not yet been completed and put into finished-goods inventory. Since each dollar's worth of inventory costs the company about 30¢ a year, a reduction in inventory will not only reduce investment, thus freeing funds, but will reduce current costs.

A good materials handling system may reduce the raw material and finished-goods inventory, but its major impact will be on the work-in-process inventory. If it can be reduced by faster and more efficient handling and by cutting the inventory needed to maintain production, the savings can be substantial.

5.5.1 Inventory Dollars

Inventory value in dollars is a measure obtainable from a good accounting system. Comparison of work-in-process dollars required by the former system with the amount required by the new system gives the reduction in inventory investment.

Usually inventory in physical units is harder to obtain and compare than one in dollars. Unless the company makes a single uniform product, it will have difficulty defining and measuring a work-in-process inventory in physical units.

5.5.2 Hidden Costs of Inventory Reduction

If inventory is fixed at too low a level, some costs will increase. These include expediting, setups for short runs, "crisis" production, and lost sales. Unfortunately, these costs are seldom identified directly but are "buried" in overhead accounts where they escape managerial attention. This contrasts to the inventory dollar total, a separate figure which management scrutinizes carefully.

5.6 PRODUCTIVITY

Improved productivity is a important management objective necessary to survival in a competitive economy. The term productivity is a general one subject to differing interpretations and perceptions. Productivity may be quantified into measures for evaluating materials handling system capabilities. Although these can be useful, the manager and system designer should be aware of possible faults due to inadequate data, to suboptimization at the expense of other functions, or to manipulation of the measure to make the system and its operation look good to top management.

5.6.1 Output per Worker-Hour

A common measure of productivity is output per worker-hour. It is the output divided by the worker-hour input. Output may be expressed in dollar value of production per unit time or in physical units processed per unit time.

5.6.2 Output per Square Foot

With the high cost of manufacturing space, efficient and productive use of facilities is imperative. Construction costs can go as high as $125 per square foot for specialized facilities. Output per square foot is a good way to measure the productivity of the facility.

5.6.3 Output per Dollar Invested

An important management goal is to maximize the productivity of the organization's capital investment. Rate of return on investment measures productivity of capital. Other factors that can be used are dollar value of the output or physical units of production relative to capital investment.

5.6.4 Output per Time Period

This measure is compatible with management thought processes. Most managers think of operations in terms of the week, month, quarter, or year. The output per period of the old and new systems can be readily compared. Also, the output per period can be compared with sales volume.

5.7 SYSTEM CONTROL

An important advantage of many automated, high-volume materials handling systems is increased control over production and inventory.

Scheduling, planning, reporting, and monitoring are better. This improvement is difficult to quantify.

5.7.1 Delays, Missed Shipping Dates

Improved production and inventory control are often a result of an improved materials handling system. This improvement may be difficult to measure quantitatively. Some characteristics, such as number of delays, expedited orders, and missed shipping dates, may be quantitatively measured even though a special study is required. Other characteristics may be real, yet intangible. Some measures may be misleading, subject to manipulation, or only remotely related to the important objectives.

5.7.2 Number of Expedited Items

The number of orders requiring expedited or rush treatment per period may be an important factor in evaluating a materials handling system design. Expedited order costs include personnel costs of the expediters, time of other personnel such as production planners and foremen when diverted from regular work, changes in the production schedules, and special setups for short runs. A good materials handling system will enable management to reduce the number of rush and expedited orders substantially.

5.7.3 Delivery Time

An important objective of the company is better customer service. One measure of customer service is the time interval between receipt of the order and shipment to the customer. This may be expressed as an average or as a statistical distribution.

In some industries, orders are placed for delivery at a future date well beyond the date required for completion of the order. In this case, the appropriate measure would be the percent of orders meeting the customer's delivery schedule.

5.8 PERSONNEL UTILIZATION

There are several criteria that measure the efficiency of personnel utilization in a materials handling system. This input factor may be part of the cost analysis or may be measured separately. Several trends increase the importance of the personnel factor:

1. The long-time commitment implied in adding a permanent employee. One firm's slogan is "Ninety days equals 30 years," or once an employee finishes the probationary period, the

company will probably have the person as an employee until retirement 30 years later.
2. Labor costs including direct pay and fringe benefits have been rising more rapidly than most other costs.
3. Many firms have experienced difficulties in hiring and retaining workers for certain undesirable or highly skilled jobs.
4. Costs per individual worker have increased greatly. This means that the employer must pay certain benefits regardless of the number of hours worked. These costs include pensions, insurance, medical and dental benefits, holidays, and vacations.

5.8.1 Number of Materials Handling Workers

A primary objective of most materials handling system design projects is reducing the number of materials handling workers required. The number of operators is easy to explain and understand and should not be difficult to determine.

5.8.2 Other

Other personnel factors that can be used are:

1. Expected reduction in labor turnover in materials handling jobs.
2. Improvement in job working conditions. This could include reduction in noise, dirt, heat, or other undesirable conditions; lighter loads to be lifted and carried; and decreased monotony.
3. Change in the skill levels required to operate the system.

REFERENCE

1. By permission. From Webster's Ninth New Collegiate Dictionary © 1986 by Merriam-Webster, Inc., publisher of the Merriam-Webster ® Dictionaries.

6

Analysis for Materials Handling System Design

6.1 INTRODUCTION

After the problem has been defined and the data gathered, the materials handling system designer begins the analysis phase. He quantifies the characteristics of the system and the requirements that must be met. If the preceding steps, problem statement and data gathering, have been carefully and thoroughly carried out, analysis is much easier.

6.2 LEVELS OF ANALYSIS

The analysis may be at one of several levels of detail and rigor. The level may be directed by management or selected by the analyst to suit the situation. The objective is to use the level of analysis that will solve the problem, yet not require more effort and expense than necessary to accomplish the mission.

Several factors enter the decision on the level of analytic rigor appropriate to the specific handling system design project.

1. The amount of money involved. A project calling for the expenditure of a million dollars warrants more time and effort than one for $10,000.
2. The time allowed for the project design. If the project design must be completed within a short period of time, the level of effort may have to be limited to meet the deadline.

3. The planning horizon. The analytical process will be shorter and simpler for a project with a short service life than for one expected to last for a long period.
4. Management direction. Management may specify the amount of detail and effort.
5. Availability of data. If the necessary data and information are available, *with accuracy and in good format*, the analysis will be simpler and quicker.

For convenience, degree of analysis detail has been divided into four levels. These levels are not discrete; they are four points selected from a continuous spectrum of difficulty. They are:

1. Highly detailed
2. Detailed
3. Moderate
4. Minimal

Let's look at these levels further.

6.2.1 Highly Detailed

A project involving large expenditures and a long expected service life will call for considerable time, effort, and expense. The design of an automobile assembly line takes a large group of engineers and technicians at least a year. A system to produce a high-volume product such as a food product, soft drink, or light bulb requires a detailed analysis to produce the required volume at a high rate of speed with minimal damage and downtime.

Many systems in this category are custom, one-of-a-kind, designed uniquely and in detail from the beginning by the company's engineering staff or by a consulting engineering firm. Each component may be individually designed for the system. Often, engineering data sheets from the equipment manufacturer are of little use since the installation may be nonstandard.

6.2.2 Detailed

An important material handling project may call for a careful, thorough analysis, yet not require the high level of detail necessary for the most important and complicated system. A conveyor system for a warehouse or a kitting, packing, and shipping line for a moderate volume product are examples of this level. The designer can use engineering data from the equipment manufacturer for much of the work. There is little sense redoing the engineering effort of the vendor's staff. However, some features, such as adjusting the system to the building and to the product, may require substantial engineering effort.

6.2.3 Moderate

Many projects require a level of effort that relies heavily on vendor's data and personnel with a moderate effort by the company's materials handling design staff. Examples are relocating a conveyor belt, increasing the height of the pallet-load stacker, or selecting a special fork-lift truck attachment. In most cases, the system is similar to previous ones; the designer can use his experience and skills to complete the design expeditiously and efficiently.

6.2.4 Minimal

By far the most numerous type of project is the small one such as a minor modification to an existing system. In this case, the designer should choose a level of analysis that will get the job done correctly with a minimal effort. In some companies, simplified drafting or even verbal instructions are used in lieu of extensive engineering drawings or sketches.

6.3 PREANALYSIS PROCEDURES

6.3.1 Problem Definition

This discussion of material handling system analysis assumes that the problem has been carefully and correctly defined. Incorrect problem definition is a common cause of wasted or ineffective design effort. Before starting the analysis, the engineer should carefully review the problem definition and statement to ensure that it is correct. A change in the problem definition may be required by changing conditions or by information developed during the data-gathering phase. The reader should refer to Chapter 2 for a more detailed discussion of problem definition.

6.3.2 Objectives

The objectives of the materials handling system may change during the project. A change in the market for the product or in the product itself may alter the system goals. New equipment or technologies may come on the market. Economic conditions may change and require a modification of the project.

6.3.3 Criteria

Before and during the analysis, the designer should review the criteria for the system to see if any should be changed. Market prospects may have changed materially. For example, a competitor's technologically advanced product may require that the company modify previously imposed constraints on capital expenditures if it wishes to remain in the market at all. On the other hand, the amount of internal and external capital available for investment may decline. This might require further constraining capital expenditures in the materials handling system.

6.3.4 Data Collection

The designer should review the data to see whether they are adequate in scope, accuracy, and quantity for the project. It is not uncommon for the designer to discover that additional data are needed. If these can be gathered before the analysis starts, time and effort will be saved and progress will be smoother.

6.4 MARKET ESTIMATION

The purpose of this section is to present the principles and techniques of market estimation by which a company can predict the market for the product in terms of the number, type, characteristics, and sales timing. This information is the starting point for the plant layout. It determines the size of the factory, the materials handling system, the equipment required, the flow arrangement, and the service facilities.

We use the term estimation rather than determination. Markets can seldom be predicted exactly—there are too many variables and uncertainties. Most companies are satisfied if the expected market volume can be predicted with ±10%.

The market estimate should include a forecast of the market trend. The sales volume of few, if any, products is static from year to year. Dynamic change is universal. Products and processes change, demand shifts, and sales rise or decline accordingly. The materials handling engineer needs to know the trend of future sales to plan for expansion and for changes.

6.4.1 Defining the Company's Market

Before the company can accurately assess its market prospects for a product, it must define that market. It may be delineated by product, by area, by type of customer, or by any factor that differentiates the company's market from other markets. Defining the market is not easy; careful, thorough study is necessary to determine what the company's market really is.

A company whose product is made of readily available local materials and is low in value compared to the shipping cost will have a market confined to a local area in which shipping costs are low enough to compete successfully. An example is building brick. Clays for this type of brick are available in many locations. The product is low in value, but heavy and costly to ship. Therefore, the company will seldom sell beyond an area measured in dozens or, at most, a few hundred of miles.

On the other hand, a company making expensive instruments will market over a much larger area. Shipping cost relative to total manufacturing cost is very small. The necessary tooling, equipment, research, and development are expensive and must be spread over a large number of units to be economically feasible. The company's market will be national and may go beyond national borders to the international marketplace.

A company may decide to serve a specialized product market. An example is custom automobile bodies. The big auto makers cannot afford to enter the custom car market; the overhead and setup expense are prohibitive. Furthermore, the customer would balk at paying the high price necessary to cover costs and make a reasonable profit. The custom auto body shop has limited overhead and is designed to be flexible and efficient in the custom auto body field. Consequently, it can make and sell its product at a price that will allow it to make a satisfactory profit. Furthermore, the customer is more willing to pay the price because he knows it is a custom shop.

6.4.2 Market Estimate as a Goal

A market estimate differs from other predictions in that it can be affected by managerial action. The chief executive officer can set goals relating to the company's desired position in the market. These may be expressed in one or more of the following terms:

1. Market share in percent of total market
2. Number of units to be sold
3. Sales dollars
4. Market dominance
5. Market position vis-à-vis a specific competitor

A definite market goal motivates company personnel to reach that goal and avoids drifting, which can easily occur when there are no specific plans or objectives. There is no static situation in modern industry; the company that tries to maintain the status quo will fall behind and eventually be absorbed into a stronger organization or it will be liquidated.

An executive may define market goals in relation to a competitor's sales. William S. Knudsen, vice-president of General Motors, was famous for his often-uttered slogan, "I vant vun for vun." His meaning was clear to all subordinates, Chevrolet sales must equal Ford's. His division did reach and then surpass his goal—to manufacture the best-selling automobile in the world.

A share of the market in percent is a frequently used goal. The manager may set a specified percent of the market as the company's target, or he may be less specific and simply say that the percent of the market must rise every year. A rising share of the market improves the company's competitive position and keeps it from falling behind.

Figure 6.1 illustrates the trend in company shares in a typical market. Rarely does the market share remain the same. It either goes down or goes up. Equally important is the trend in the total market. An increasing share of an expanding market is characteristic of a prosperous company. An increasing share of a decreasing market is a signal to look for new markets and new products. These situations are illustrated in Figures 6.2 and 6.3.

Market dominance is harder to define. The company that is dominant in a market controls many facets of that market, such as product characteristics, prices, and marketing strategies. Usually the company with the largest market share is dominant, but this is not necessarily true.

An example of market dominance is the Bell Telephone System. Phone transmission equipment must be compatible with Bell's; otherwise the independent phone company and its subscribers are isolated from other phone systems. Accounting and billing must also be compatible. Dominance extends even to the color of the advertising pages. Nationwide they are yellow, although there is nothing to prevent an independent from printing its advertising pages on green or blue paper.

6.4.3 Units for Expressing Market Estimates

The most useful market parameter for materials handling and plant layout design is the number of units. This is a physical quantity which can be the basis for calculating product flows, equipment, storage, and other requirements. Few plants make only one product; much commoner is the factory that manufactures a number of different products in one facility. The different products may be a family of similar products, such as containers of different sizes, or may be a combination of different products using the same production processing equipment.

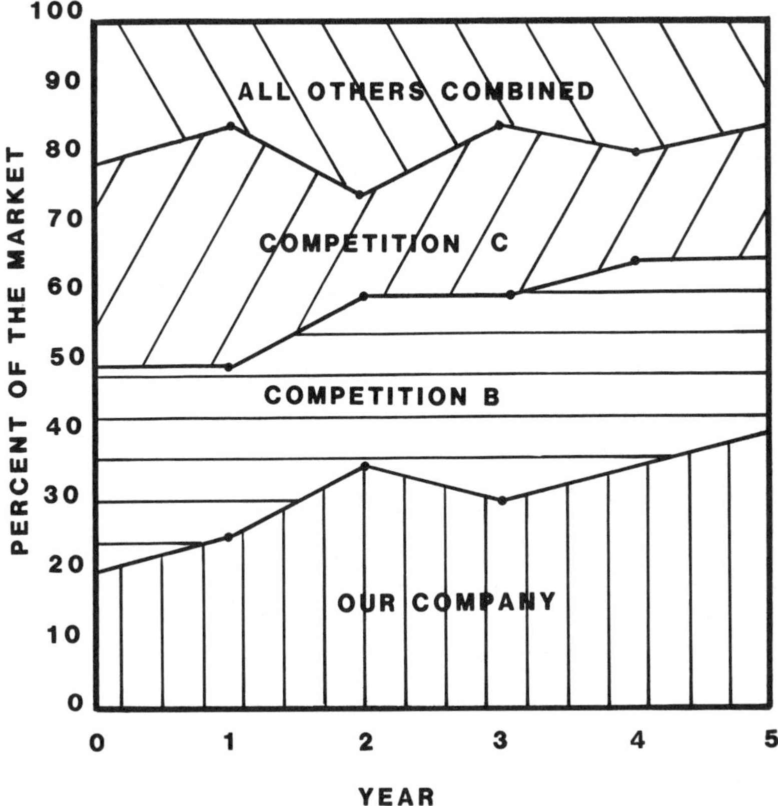

Fig. 6.1 Share of market in percent.

For financial planning and ease of understanding, a company may express its sales goals in total sales dollars. This approach is useful for financial budgeting and planning for the various resources necessary to make the company's products. It also serves as a common denominator for combining estimates of a number of different products into a single figure. However, it is not suitable for materials handling system design.

6.4.4 Methods for Estimating the Market for a Product

There are several methods available for estimating the company's market for a product. The market research department may use any one or a combination, depending on the company's resources, the value/cost relationship of the estimate, the availability of data, and

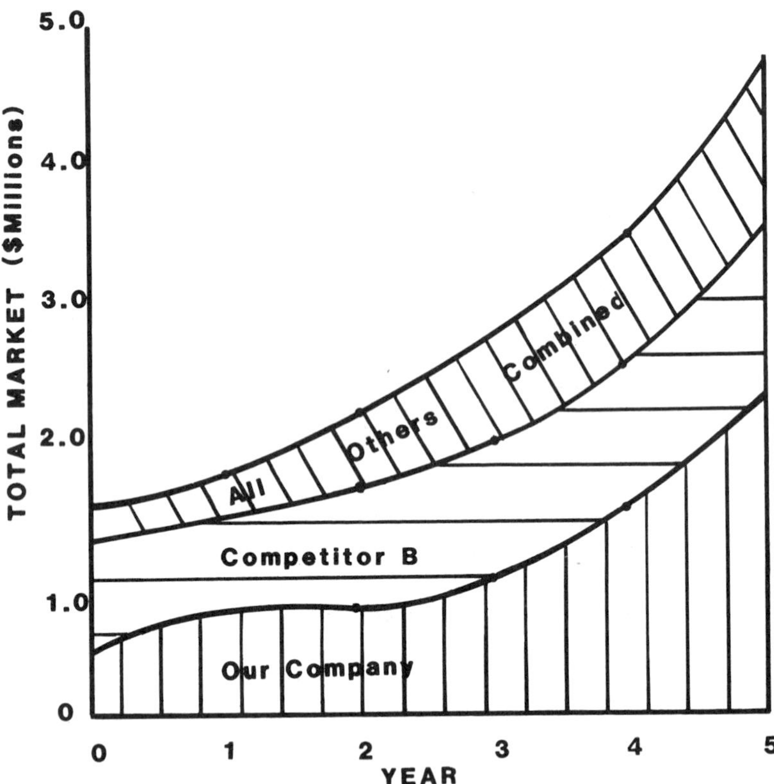

Fig. 6.2 Share of an expanding market by dollar volume.

managerial personal preference. The methods of market estimation generally used are:

1. A sampling survey
2. A known relationship to a widely published estimate or trend
3. Feedback from sales and service representatives and similar sources
4. Managerial judgment including extrasensory perception
5. Extrapolation of past company sales trends modified by economic, social, and technological predictions

A sampling survey is a question poll of a scientifically chosen sample of the market. If the questions are correctly phrased so that the answers are unbiased, if the poll-taking organization conducts its survey without bias, and if the sample accurately reflects

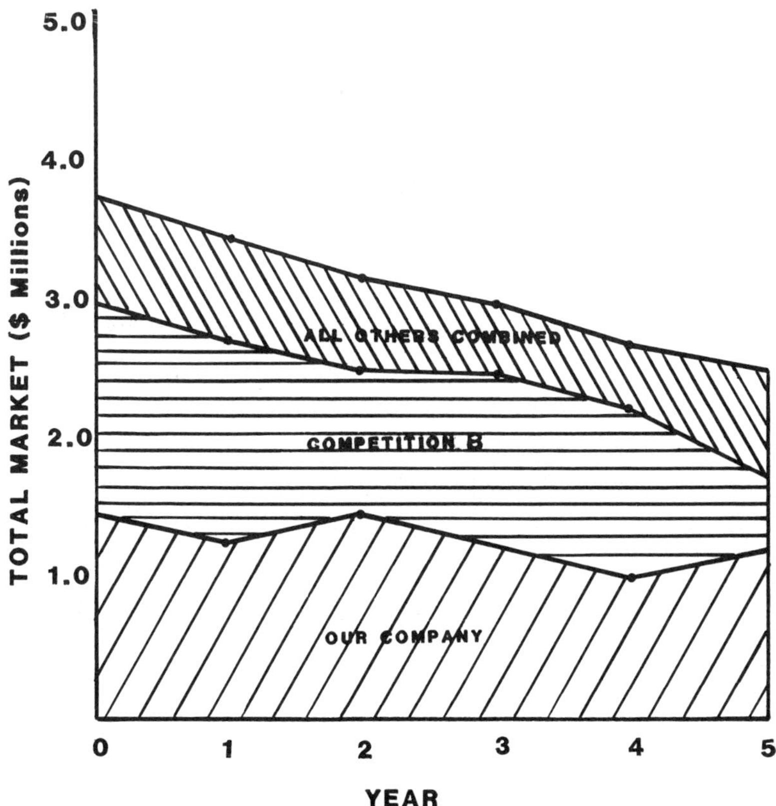

Fig. 6.3 Share of a declining market by dollar volume.

the company's market, the results will provide a good estimate of the company's market. This method is frequently used by large consumer-goods manufacturers who poll samples of the consuming public. They usually engage specialized firms for the study.

The market survey may be subject to bias in the questions themselves or in the way the pollster presents them to the respondent. A famous example of a biased question occurred in a survey taken for an automobile manufacturer in the late forties. The respondent was asked what characteristics he wanted in a new automobile. Many replied that they wanted gas economy, safety, reliability, and ease of maintenance in preference to appearance and high horsepower. Actual sales showed that appearance and horsepower were the overwhelming motivations of the new car buyers. Either the respondents answered what they thought they should or gave the answers that they thought the pollsters wanted. The problem of bias is more common among low-income families, who tend to be suspicious of outsiders and to be guarded and noncommunicative in their answers.

If a company's product bears a well-defined relationship to a readily available economic statistic or published trend, this is a good basis for a market estimate. One example is building hardware. If the company's sales correlate well with the number of residential building permits issued, the expected sales can be derived from predictions of the number of housing starts. However, the company's marketing department should be alert to changes in the relationship. In the building materials field, the growth of do-it-yourself has changed the market for some building materials and components and consequently market prediction methods.

Sales and service representatives of the company provide useful information on the company's products and sales prospects. They can be alert to important facts such as the customer's reaction to competitive new products, deficiencies in the present line, and improvements wanted by the customers. This is particularly important in the capital goods market, where the customer has definite needs and desires and is technically competent to evaluate the company's and competitor's products.

Managerial judgment varies from careful analysis and evaluation of available data to a wild hunch based on little data and less thought. A good manager can frequently arrive at a surprisingly accurate estimate of the company's market potential. Some engineering research indicates that a few managers have the extrasensory ability to estimate accurately without being able to identify the specific data or thought processes that were used to arrive at the estimate. A small company in which the managers are in constant contact with customers often has an advantage over a large one in assessing the market and moving accordingly. The key managers of a large organization are often out of direct touch with the customers and must rely on second- or third-hand information. They also face a time lag in receiving and acting on current market information.

A change in manufacturing technology may alter a market. Some beverage manufacturers have lost competitive position because their volume did not justify the automation necessary for survival against highly automated rivals.

New materials may change a market substantially. Synthetic fibers replaced natural ones in clothing for a time. However, the natural fibers have made a comeback because cotton and wool have wear and appearance characteristics that synthetics have yet been unable to duplicate.

Errors in estimates due to substantial changes in the market are harder to predict and more severe in their influence. An example of a technological change that completely restructured a market is the dry copier Xerox®, which completely took over the copier market and drove wet copiers into oblivion.

6.4.5 Tolerance of Market Estimates

A market estimate is not a precise prediction of the market volume. It is an estimate of the future subject to mistakes in estimating procedures and data, to changes in the market or technology, and to chance causes. A market estimate should have a statement of the possible error.

One source is the statistical error involved in sampling. Most polling organizations will survey about 300 persons or households in a locality for a study. The sampling error due to random causes for this number of observations is ±5%, with a 95% probability. Three hundred observations is a good balance between the cost of polling, which is closely related to the number of persons polled, and the accuracy of the estimate, which is proportional to the square root of the number in the sample. To double the accuracy (halve the error estimate) the pollster must quadruple the number of observations in the sample.

The company's market volume is the product of two components, the total market for the product and the company's share of that market. One method used for estimating total future market volume is to sample sales of the item in a small area. This number is then extrapolated to the entire region or nation. For example, if the annual sales of an item in a market area of 100,000 persons were 55 units, we could estimate an annual national sales volume of 12,100 units, based on the total United States population of 220,000,000 persons. The market researcher should be cautious that the community sampled is representative of the entire market. Frequently, several communities in different regions and with varying characteristics will be sampled to arrive at a better estimate.

The company's share of the total market is much harder to estimate. Estimation of present market share can be obtained by comparing company sales to total market. For example, if the total market is 450,000 units per year and our company sales are 115,000 units per year, we have a 25.6% share of the total market. Another way of estimating market share is to include questions on brand or manufacturer in the sampling survey used to determine total market. In either case, the data are estimates and should be regarded as such.

In some industries such as automobiles, market data by manufacturer and make are freely available from published sources. This makes the job of estimating present market easier. However, this source of data provides negligible information on competitors' plans or future trends.

Market share depends in part on the results of managerial decisions and actions. The executive must decide the price and the product characteristics and estimate what share of the market can be captured. Penetrating an existing market is harder than establishing one for a new product. If the company is already into the market, it is easier to estimate what the company's share is likely to be. This can be based on trends, aggressive policies, competitor's expected actions, and present market share.

A company may predict its market volume as an extrapolation from past sales. If sales have risen steadily in the past few years, the market research department may assume that next year's sales will follow this pattern. This is satisfactory when there are no substantial changes in the market and when product sales are correlated with rising income and population.

This method assumes that past trends will continue unchanged into the future. The market research director should recognize this assumption and carefully evaluate its validity. New processes, new products, a shift in consumer demand, or an innovative competitor may ruin this assumption and lead to unwarranted, even disastrous, market predictions.

6.4.6 Trend Forecasting

It is not enough for you as plant layout designer to have an accurate estimate of the market for the coming year. It is important to have some estimate of the trend for at least 5 years into the future. Many decisions, such as plant location, original plant size, and provision for expansion, depend on long-range future trends as well as on immediate sales prospects.

A basic fact of long-range forecasting is that numerical estimates are subject to increasing error and adjustment as the time horizon moves farther into the future. A commonly used pattern is:

Years into future	
1	Fairly definite for planning purposes
2–4	Satisfactory for contingency planning
5	Limit of definitive planning
10	Limit of future planning

6.4.7 Style Goods

Some goods, such as women's dresses and men's neckties, follow a definite cycle in sales. A new item moves slowly at first. As it becomes popular, sales increase rapidly. After a short while, interest in the item begins to decline and sales drop off. At the end of the cycle, no one will pay full retail price and the remainders must be sold at a substantial discount. The entire cycle may take only a few months.

In a style goods business, predicting the market and producing to demand are all important to the company's profitability. The manager must decide what styles will be popular in the near future; he cannot wait until the style becomes popular. If he waits for definite popularity signals, he cannot design the item, purchase the materials, get into production, and deliver the goods in time to have satisfactory sales. In the later phase of the cycle, he must begin cutting back in time to avoid getting caught with a large inventory of outmoded merchandise. Style merchandisers and manufacturers were among the first to use real-time computers to keep abreast of current sales and inventories.

6.5 VOLUME ESTIMATION

Key parameters in material handling system design are the characteristics of the material and the volume to be handled. The engineer may be given this information as part of his assignment, or he may have to gather the data himself.

If the parameters are given, the engineer should take two steps:

1. Verify the correctness and adequacy of the data for his purpose. Sometimes the original data are incomplete and must be augmented. The time and effort will depend on the adequacy of the original data.
2. If the system involves present product, equipment, and facilities, the engineer should make an observation tour. Many features will show up on the floor which may be unclear or missing from drawings, specifications, or project writeups.

6.5.1 Standard Materials Handling Unit

A standard materials handling unit is defined as a predetermined number and arrangement of product which can be moved as a single item. A common unit load is a pallet loaded with a specified number of cartons of consumer goods in a standard arrangement. The unit load is assembled at the end of the manufacturing line and moves as a single item through finished goods storage, shipping, and transportation, ending with receipt at final destination. Then it is broken down for distribution to the retail outlets. See Chapter 4 for further discussion of unit loads.

The engineer must determine the material unit to be moved. In some organizations, this may be determined by the product, existing equipment and practices, regulation, or transportation rules. In others the materials handling system designer may have some latitude in determining the most efficient unit for his situation.

6.5.2 Product Mix

Product mix is defined as the number and proportion of different items the factory produces. It may be determined by marketing as well as production consideration. Normally, a few items will account for the bulk of sales volumes. The company may produce the slower-moving items in order to offer a complete line to its customers.

6.5.3 Mixed Materials Handling Units

Unfortunately, most materials handling systems must accommodate a variety of units. A factory may produce many different parts, sizes, and assemblies within a single manufacturing system. The engineer must design his system to handle what the factory uses and produces. He must also be prepared to modify the system to handle new items and different product mixes.

6.5.4 Limits

The materials handling unit may be limited in size, weight, or other characteristics:

1. Size. Transportation or space availability may limit the unit options. Examples are:
   a. Pallet loads are normally limited to standard pallet sizes. Although it may be technically feasible to specify a non-standard pallet, there will be transportation difficulties and extra costs that will overwhelm savings.
   b. The 55-gallon drum is standard for many liquids and dry powders.
   c. Building clearances may limit load dimensions. Building alterations may be structurally infeasible or excessively costly.
2. Weight. The materials handling unit may be limited by weight.
   a. If the load must be manually handled somewhere in the system, the maximum weight may be specified by law, regulation, union contract, fatigue, or custom.
   b. Floor load limits may restrict total weight.
   c. Equipment strength may limit weight.

3. Cost considerations may influence unit load. For example, it is usually cheaper to use a standard, off-the-shelf container than a custom-designed one. In addition to being more costly, custom equipment may be harder to service and maintain.

6.6 VOLUME ANALYSIS

A major parameter of the materials handling system is the volume of product that must be moved. This is the prime determinant of the flow rate, the system concept, the type of equipment, and the size of the system. This is particularly true of an automated system since it has less flexibility to meet overloads and demand variations.

6.6.1 Basic Volume Unit

The first volume parameter to determine is the unit of measurement. If the system operates with a standard unit or has only a single product, the standard unit is already settled. If not, the product that the system is to move must be examined carefully to determine the best standard move unit.

Installation of new materials handling system may be a good time to introduce a new standard unit into the system. A standard pallet load may be used in place of a variety of bins and tote boxes. An automated system operates on the basic design concept of a standard load unit with a standard load carrier.

Pallets

A standard pallet with load is a useful volume measure for many systems. These may be either 40 by 48 or 48 by 48 in. One advantage of the standard pallet is that different amounts of different products can be loaded on a pallet and moved as a unit. A pallet may be used as a base for a bin or a number of nested tote boxes.

Some automated systems use slave pallets. These are pallets that are an integral part of the system and do not leave the system for manufacturing operations, storage, or order picking.

Product

Another useful measure of volume is the product. For example, an automobile or appliance assembly line is designed with one product item as the unit of movement. A special fixture or base on which the product is assembled is the basic unit of system design.

A recent development is to make an automatic guided vehicle (AGV) the basic move unit for an assembly line. The fixture is mounted on the vehicle and moves along the preplanned path with programmed stops and variable speeds. AGV systems are discussed further in Chapter 17.

Carts, Trucks, or Carriers

A distribution center or warehouse may use a four-wheel cart as the basic move unit. The cart may be loaded with several different products in varying quantities, yet the uniform unit of move through the automated or semiautomated system is the cart. An overhead conveyor system may use a carrier as the move unit.

Tote Box

If the system is such that a standard unit such as a pallet is too large, the tote box or smaller container may be best. This would be the case if the individual items are quite heavy or the system calls for some manual intervention. An example is order picking, kitting, and packaging orders at a distribution center for shipment to retailers.

Bulk Product

Some materials handling systems are devoted to a single bulk product, such as coal or grain. In this case, the standard move unit is the one used throughout the industry. For example, coal throughput is measured in tons and grain in bushels or tons. Some systems handling bulk goods for export use the metric ton instead of the traditional English ton.

Combination

Many systems call for a different move unit in different parts of the system. A food manufacturer may move the product in bulk through the basic processing, into individual packages, and then into unit pallet loads for finished-goods inventory and shipment. If the system must handle more than one move unit, data must be obtained separately for each segment through which the product moves.

6.6.2 Time Unit

A rate must be expressed in units per time interval. The usual time interval for expressing volume rates through a materials handling system is the hour. Shorter intervals are too much affected by fluctuations and by the time per individual transaction. Longer intervals are harder to use and may mask important hour-to-hour variations.

6.6.3 Uptime of Interfacing Systems

Uptime is the percent of time a system is operating. No system can operate 100% of the time. Here we are not concerned with the uptime of the materials handling system itself. This will be determined in the system design phase of the project. We are looking at the reliability of systems that interface with the materials handling system.

6.6.4 Variability

Another parameter is the variability of the flow. Average flow is inadequate if the hour-to-hour volume may vary from −50% to +100% of the average volume. This variability may occur on a regular basis, such as little or no movement immediately after starting time or during a product changeover or random machine breakdown.

Very few company records show this variability. It can be obtained through a one-time, special-observation study. Work sampling is a good technique. In a work-sampling study, an observer takes data at predetermined random intervals throughout the day. From these, he can develop a frequency distribution of volume rates. If the study is carefully designed, the results will be accurate enough for the system designer.

6.6.5 Volume Calculation—Example

A food manufacturer is planning to install an automated materials handling system to palletize the cartons and move them to shipping or to finished goods storage. The following data are presented:

Monthly volume—1,500,000	
Cartons per pallet (unit load)	33 cartons
Working hours per month	335 hr
Uptime of system through packing into cartons	94%
Ratio of peak load to average load when system is operating	1.0 to 1.0
Ratio of minimum load to average load during operation	1.0 to 1.0

The calculation of volume needed is

$$\text{Pallets per month} = \frac{1{,}500{,}000}{33} = 45{,}455 \text{ pallets/month}$$

$$\text{Pallets per hour at 100\%} = \frac{45{,}455}{335} = 136 \text{ pallets/hour at 100\% uptime}$$

$$\text{Volume variations during operation} = \text{None}$$

$$\text{Volume for system design} = \frac{136}{0.94} = 145 \text{ pallets/hour}$$

6.6.6 Volume Adjustment for Scrap Loss and Efficiency

There are two reasons a factory cannot attain theoretical production rates—scrap losses and efficiency. Scrap loss is the percent of production that must be discarded as not meeting specifications. It may include material losses unavoidable in the process. For example, an automatic screw machine cutting small fasteners from bar stock cannot use the last few inches of the bar which hold the stock in the machine. Scrap losses may be substantial. In transistor manufacturing, scrap losses may reach 90%; 10 transistors must be started into production to produce one good one. Scrap losses are expressed in percent of input production. For example, if an average of 95 good pieces are produced for every 100 started into the process, the scrap rate is 5%.

Efficiency is normally expressed as a percent of theoretical maximum efficiency. One hundred percent cannot be attained because inevitably there will be machine breakdowns, external or internal delivery delays, personal and scheduled breaks, or other problems. In some industries efficiency is expressed in productive minutes per hour. In machine shops, the rule of thumb is 50 min of production per elapsed hour of work time.

The industrial engineer must allow for both scrap losses and efficiency losses in planning for his layout. The formula for scrap loss is:

$$\text{Number of units to be processed} = \frac{\text{Number of good units required}}{1 - \text{scrap loss percent}}$$

The formula for efficiency is:

$$\text{Total time for planning} = \frac{\text{Theoretical or standard time}}{\text{Efficiency percent}}$$

Example

Given

Number of parts required annually = 1000
Average scrap loss = 8%
Standard time per part = 3.65 min
Efficiency = 90%

Required

Number of hours per year for planning capacity

Solution

$$\text{Number of parts to be processed} = \frac{10{,}000}{1 - 0.08} = \frac{10{,}000}{0.92}$$
$$= 10{,}870 \text{ units}$$

$$\text{Theoretical standard hours per year} = 10{,}870 \times 3.65 \times \frac{1}{60}$$
$$= 661.3 \text{ hr/year}$$
$$\text{Planning hours per year} = \frac{661.3}{0.90}$$
$$= 735 \text{ hr/year}$$

6.7 DEVELOPING ALTERNATIVES

Searching out and developing feasible alternatives to accomplish the task is a key factor in successful system design. This requires experience, judgment, and balance. Time and its cost in money are never unlimited; all possible alternatives cannot be discovered and investigated. On the other hand, it is important not to miss an alternative that could be the best.

6.7.1 Sources of Ideas

Where does the systems designer get his ideas? Few, if any, ideas are original. They are borrowed from other organizations, applications, and sources. They may be modified, combined, abridged, or otherwise changed for the new system. Important idea sources for improvement are new products and new technologies. They may make possible methods which are desirable but were previously infeasible.

Not all new products and technologies are applicable or successful. Industrial folklore is full of new ideas and applications that failed in practical operation. The old saying is good advice, "Be not the first by whom the new is tried nor yet the last to lay the old aside." The designer's toughest task may be to sort out the good ideas from the rest.

Experience

The systems designer accumulates knowledge and experience as he moves along in his career. He sees what works and what doesn't. From this, he can develop useful alternatives and discard those which hold little or no promise of success.

Experience can be a handicap. Most persons stick with methods that have worked for them in the past. They are more comfortable with the known. There is less risk in recommending a known solution than in advocating a new, untried approach.

Trade Literature

The materials handling systems designer should subscribe to periodicals both in materials handling and in his own industry.

Trade publications are a good source of new ideas and concepts, particularly those involving new and improved products. There are articles on interesting applications, and short features on trade happenings and new products. Ads feature new products and improvements.

Trade publications should be read with some degree of skepticism. Not all products, installations, or ideas are successful. Failures are rarely written up; have you ever seen an article titled "How Our Million-Dollar Automated System Failed and Had to Be Torn Out"?

Trade Meetings

If the time and expense of the materials handling system and the company's resources warrant, a trip to a materials handling trade show will be worthwhile. The designer can see and inspect new products, ask questions, and exchange observations with other attendees. One of the biggest is the annual materials handling show. Others include specialized shows such as the annual Automated Storage and Retrieval System Meeting and those devoted to a single industry such as steel, plastics, or grocery warehousing.

Other Persons

Other persons in the organization may have good suggestions. They may be in other plants or in other functional departments such as plant engineering. Hourly workers are an untapped source. Today's worker, particularly in the skilled trades, may have good ideas based on experience working with similar systems.

Sales Representatives

Sales representatives of materials handling equipment manufacturers may be a good source of new ideas. They are familiar with their company's products and trained in applications to customers' problems. They can provide good information even though their job is promoting their employers' products.

6.7.2 Preliminary Selection

One of the most important, yet least understood, steps in the design process is the preliminary selection of a few alternatives for further study and analysis. Seldom can much time and effort be devoted to even a brief analysis of all possible alternatives. On the other hand, the best alternative may be discarded at preliminary stages and not considered further. The folklore of American industry abounds with stories of immensely profitable opportunities passed over in the preliminary selection process.

There are no mathematical algorithms for selecting the most promising alternatives in the preliminary sort. Experience and good judgment seem to be the factors most important in making a good choice. Again, folklore relates instances in which an experienced decision maker with a good record makes a serious error in selecting alternatives for analysis and study.

One rule of thumb is that the number of alternatives chosen for further study should not exceed three to five. This strikes a balance between the possibility of omitting an optimum alternative against the cost and time of examining many alternatives in detail.

6.8 FLOW PROCESS CHART

6.8.1 Introduction

Good analysis calls for a method of organizing facts in a rational and easily visualized manner. The flow chart is such a tool. It lists the functions required to reach the desired result, which may be a completed part, a component or subassembly, a materials movement, or an office procedure. Flow charts of present and proposed methods can be compared to evaluate improvements.

Several different flow charts formats exist. The simplest is a listing of the required steps. Other forms can be quite elaborate. The company may prescribe a particular form and furnish blanks. In small organizations, the analyst may be free to follow his or her own preferences. A sample flow process chart form appears in Figure 6.4.

6.8.2 Construction of a Flow Process Chart

In this section, we shall follow a worker as he is order-picking two items. This will show how the chart is prepared and what information is included.

Heading Information

The heading contains information needed to identify the process charted and the persons involved. It should include the following:

1. Title or description of the process being charted
2. Chart reference number
3. The name of the person preparing the chart
4. The department or place of the process
5. Drawing, specification, or job description references
6. Whether chart is present or proposed method
7. Beginning and ending part of the process

FLOW PROCESS CHART

PRODUCT ______________________________

COMMENTS ______________________________

REFERENCES ______________________________

STARTING POINT ______________ ENDING POINT ______________

PROJECT NO. ______________ PREPARED BY ________ DATE ______

QUANTITY	DISTANCE	TIME PER UNIT	SYMBOL	DESCRIPTION

Fig. 6.4 Flow process chart form.

8. Page number and number of pages if there is more than one page
9. Worker's name
10. Machine, part, and operation number references
11. Remarks or information not covered by standard form
12. Date chart was prepared

The company may use a preprinted form which omits some items. Some processes may not call for all the above information. For example, a machine and part number would not be entered for the order-picking operation.

Body of the Chart

The body of the chart has five columns: quantity, distance, description, time per unit, and symbol. If the information to complete the quantity and time per unit columns is unavailable or irrelevant, these columns can be left blank. Distance is entered only when there is travel from one place to another.

Each step is listed in sequence. The description should be complete enough for the reader to follow, yet as brief as possible. The standard symbols used are illustrated in Figure 6.5. These symbols may vary from company to company. In the early days of industrial engineering, long lists with many different symbols were developed. Experience has shown that the short list of five symbols is easier to use and understand.

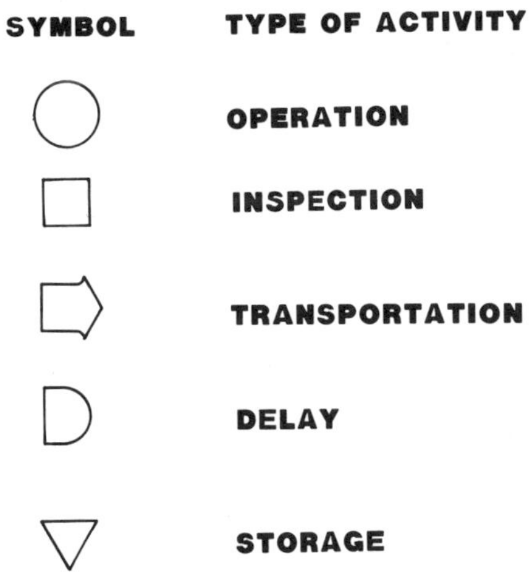

SYMBOL	TYPE OF ACTIVITY
○	OPERATION
□	INSPECTION
⇨	TRANSPORTATION
D	DELAY
▽	STORAGE

Fig. 6.5 Symbols for flow process charting.

Summary

The summary shows the total number of entries in each category and the number of feet traveled. The summary is useful in looking for ways to improve the process. There is an implicit assumption that reducing the number of steps in a category will improve productivity. This may not be true in all cases, but is a good working guide. A similar assumption is that reducing the number of feet traveled will also improve productivity.

Following the Flow

The flow process chart is a versatile and effective tool for the materials handling system designer and analyst. It can be used to follow the actions of an individual worker or piece of equipment. In our example, the flow followed the worker filling an order. It could follow the path of one of the items in the order. The subject can be a part passing through manufacturing from receipt of raw material to delivery to final assembly or stores. Another use of the flow process chart is following the path of a piece of information in the form of either a document or a signal.

Sources of Information

The analyst obtains flow chart information from several sources. The most important is personal observation of the process. The observer can pick up facts and ideas from direct contact which would be missed if only data from second-hand sources were used. Sometimes, information will come from the worker involved. This is important when the flow varies due to load fluctuations, downtime, or other causes. The worker's supervisor may also provide information. Written records such as job descriptions may be useful.

The time and effort the analyst expends will be dependent on the importance of the job. If a process is repeated many times by several workers, thorough examination may be warranted. On the other hand, a process involving small amounts of money will justify only a limited effort.

6.8.3 Using the Chart

The purpose of a flow process chart is to improve the productivity of the materials handling function. Otherwise, there is little point expending the time and effort. The chart helps the engineer's thoughts and examination by organizing the process details for easy analysis.

Each engineer will go about the examination in his or her own way. It is an individual process which each person develops through experience and education. However, certain general principles can be applied to improve the efficiency and results of the analysis.

Look for Waste

The first step is to look for waste, defined as unnecessary or unproductive expenditure of time, effort, material, or money. Perhaps the whole process is unproductive. If so, eliminating it will save 100%, the maximum possible cost reduction. The analyst should question the necessity of each motion, each step, and each transportation and eliminate those which are unnecessary. After eliminating the unproductive, the analyst should look to see which steps are being performed inefficiently. If there is a better way, it should replace the inefficient one. Perhaps a combination of one or more steps will improve the process.

Question Checklists

Many analysts use checklists of questions to ask, points to consider, inefficiencies to look for. One example of a short and simple list is included here. The analyst should decide what list best fits his or her work style and which will give the best results.

The short, simple, yet popular checklist is:

1. What can be eliminated?
2. What can be combined?
3. Can the sequence be changed?
4. What can be improved?

This checklist is applied to the flow process chart example later in this chapter.

Some dangers exist in using a checklist. One is that the analyst may be tempted to rush through the list without really thinking about each item. If he does so, he may miss an important improvement. Another danger is that the list may not fit the job at hand. Particularly if the list is short, it may omit key operations. If the analyst is not alert to important questions not on his list, he may miss a key improvement.

Before and After Charts

The analyst may prepare two flow charts for a simple process, one showing the present method and one the proposed. The before and after charts can be compared to show the reduction in the number of operations, inspections, moves, storages, and distances. This approach makes the assumption that a reduction in the number of entries in each category will reduce the overall cost of the process and improve productivity.

6.8.4 Examples

In this section, we shall present a few examples of flow process charting. These are relatively simple to illustrate the technique quickly to the reader. A long, complex process can be charted using the same approach. The analysis and improvement is longer and will take more time. Other examples can be found in texts and books on methods improvement and work measurement as well as in the current literature.

A Materials Handling Example

The first example is a simple order picking of two items. In the original method, the worker got the order document, walked to the bin, picked the first item, brought it to a packaging table, walked to the second bin, picked the second item, back to the table, checked the order, and carried the tote box to the shipping department. This is illustrated in Figure 6.6.

Several possible improvements suggest themselves. A carousel would eliminate most of the walking. An automated miniload transporter system would also eliminate the walking and probably reduce the cycle time per pick. However, both these alternatives require capital investment. Another method not calling for capital investment is for the picker to move from the first bin directly to the second while carrying the first item manually. This revised method is shown in Figure 6.7. The summary shows a reduction in average travel time.

Order picking a variety of items from different locations will have considerable variation in distances traveled. In this situation, an average travel distance is used. If the job warrants, the analyst may take enough observations to get a frequency distribution of travel distances. This may be useful if there is a problem with variable order-waiting queue length.

A Product Example

Figure 6.8 is the engineering drawing of a typical flange used in large volume. The original method called for a machined sand casting with cored inside diameters. This required facing the base, boring and reaming the two inside diameters, and drilling four holes. The flow process chart is shown in Figure 6.9. The manufacturing process was costly and time-consuming. Various ways to reduce the manufacturing cost were investigated.

An examination of service conditions disclosed that a zinc die casting would meet all requirements. The outside dimensions, the finish, and the two inside diameters were satisfactory in the "as cast" condition. The four holes could not be produced in the die casting because of limitations of the dies and the die casting process

PAGE 1 OF 1

FLOW PROCESS CHART

PRODUCT Order Picking Two Items

COMMENTS Warehouse #2. Present Method

REFERENCES Job Class 6B

STARTING POINT Empty Workplace **ENDING POINT** Empty Workplace

PROJECT NO. 732-A **PREPARED BY** ____ **DATE** 2/13/84

QUANTITY	DISTANCE	TIME PER UNIT	SYMBOL	DESCRIPTION
	10'		○	Get empty tote box and put on table
			○	Pick up and read order
	40'		⇨	Walk to Item #1 bin
			○	Pick required number of Item #1
	40'		⇨	Walk back to table
			○	Put Item #1 in tote
	32'		⇨	Walk to Item #2 bin
			○	Pick required quantity of Item #2
	32'		⇨	Walk back to table
			○	Put Item #2 in tote
			□	Verify accuracy of order filling
	100'		⇨	Carry tote box to shipping
			○	Deliver tote to shipping clerk
	100'		⇨	Walk back to table
				Summary
				Operations 7
				Transports 6 Inspections 1
				Feet traveled 454'

Fig. 6.6 A flow process chart of order picking—original.

FLOW PROCESS CHART PAGE 1 OF 1

PRODUCT Order picking two items - Revised Method
COMMENTS Warehouse #2
REFERENCES Job class 6B
STARTING POINT Empty Workplace ENDING POINT Empty Workplace
PROJECT NO. 732-A PREPARED BY R.M.E. DATE 2/13/84

QUANTITY	DISTANCE	TIME PER UNIT	SYMBOL	DESCRIPTION			
	10'		○	Get empty tote box and put on table			
			○	Pick up and read order			
	30'		⇨	Walk to Item #1 bin			
			○	Pick required number of Item #1			
	25'		⇨	Walk to Item #2 bin			
			○	Pick required number of Item #2			
	30'		⇨	Walk back to table			
			○	Put items #1 and #2 in tote box			
			□	Verify accuracy of order filling			
	100'		⇨	Carry tote box to shipping			
			○	Deliver tote box to shipping clerk			
	100'		⇨	Walk back to table			
				Summary			
					Original	Revised	Improvement
				Operations	7	6	1
				Transport	6	5	1
				Inspection	1	1	1
				Distance	454'	295'	159'

Fig. 6.7 A flow process chart of order picking—revised.

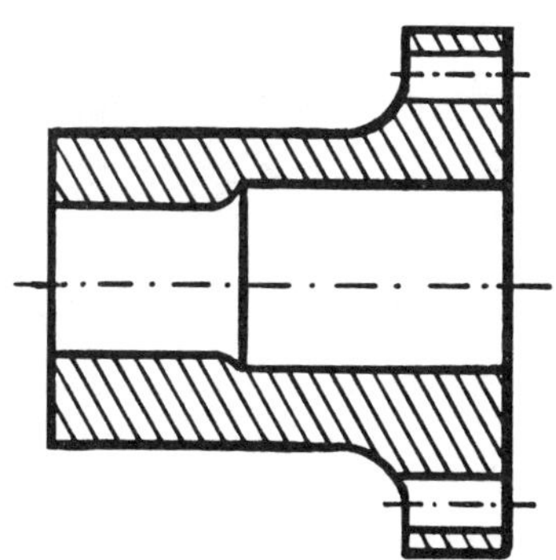

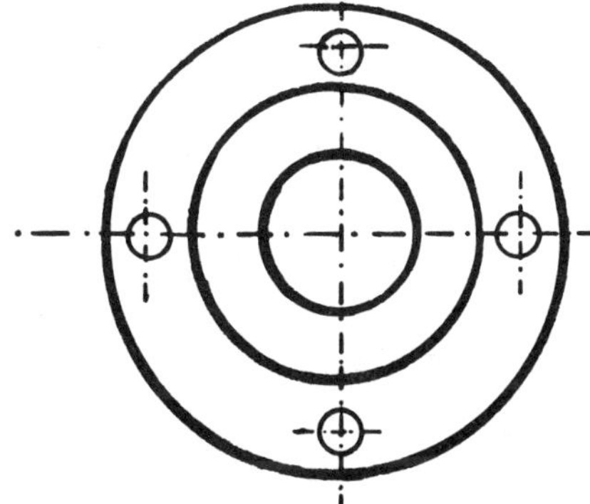

Fig. 6.8 A flange.

FLOW PROCESS CHART PAGE 1 OF 1

PRODUCT Flange
COMMENT Machined sand casting - original method
REFERENCES Drawing 6.8-5
STARTING POINT Unfinished casting ENDING POINT Stores
PROJECT NO. 6.8.5A PREPARED BY RME DATE 4-11-84

QUANTITY	DISTANCE	TIME PER UNIT	SYMBOL	DESCRIPTION
	20'		⇨	To turret lathe
			○	Insert in fixture
			○	Face base
			○	Bore small diameter
			○	Bore large diameter
			○	Ream small diameter
			○	Ream large diameter
			○	Remove part from fixture
	15'		⇨	To drill press
			○	Drill four holes
	30'		⇨	To inspection
			□	Inspection
	40'		⇨	Stores
				Summary
				Operations 8
				Transportations 4
				Inspection 1
				Distance 105'

Fig. 6.9 Flow process chart for machining flange—original.

FLOW PROCESS CHART PAGE OF

PRODUCT Flange

COMMENTS Die casting - improved method

REFERENCES Drawing 6.8-5

STARTING POINT Die Casting **ENDING POINT** Stores

PROJECT NO. 6.8-5 B **PREPARED BY** R.M.E. **DATE** 4-11-84

QUANTITY	DISTANCE	TIME UNIT	SYMBOL	DESCRIPTION			
	25'		⇨	To drill press			
			○	Drill four holes			
	30'		⇨	To inspection			
			□	Inspection			
	40'		⇨	To stores			
				Summary			
					Original	Improved	Difference
				Operations	8	1	7
				Transportations	4	3	1
				Inspection	1	1	0
				Distance	105'	95'	10'

Fig. 6.10 Flow process chart for machining flange—revised.

itself. The revised method is shown in Figure 6.10 along with a summary of savings in the number of operations and the distance moved.

6.9 OPERATION PROCESS CHART

An operation process chart is a method for graphically presenting for analysis the steps required to make an assembly or subassembly or to complete a process or procedure. An example is shown in Figure 6.11.

If the assembly contains a large number of parts and operations, it should be broken down into a main chart and subassembly or sub-operation charts. Too much detail on a chart takes too much space and confuses the viewer. The complete operation chart for an automobile or large appliance would cover the wall of a large room. This is too overwhelming to be useful.

6.9.1 Construction of the Operation Process Chart

The operation process chart combines a flow process chart for each component of an assembly with a tree diagram of the sequence in which the components are assembled into the final product. The symbols used are the same as those for the flow process chart; see Figure 6.5. The column at the right is used for the first component, usually the base upon which other components are assembled. Each successive component occupies a column to the left of the base column. The columns from right to left are in the order in which the components are assembled to the base. Crossover lines should be avoided or minimized. Purchased components upon which no work is done may be introduced by a straight line from the left.

The operation process chart should show only enough detail for good analysis. If too much detail is shown, the chart is difficult to follow and use. A subassembly may be analyzed by a separate operation process chart which is introduced into the main operation process chart as a single item.

Travel distance in feet can be added to the operation process chart if desired. In order picking, travel distance is a key factor and is included. If travel distance is not important, it can be omitted. Operation times can also be included. However, the chartist should keep the chart as simple as feasible to accomplish its mission. Unnecessary detail and complexity interfere with efficient use of the chart.

6.9.2 Example—Order Picking

The operation process chart for order-picking two items manually is shown in Figure 6.12. Unnecessary travel is highlighted. The

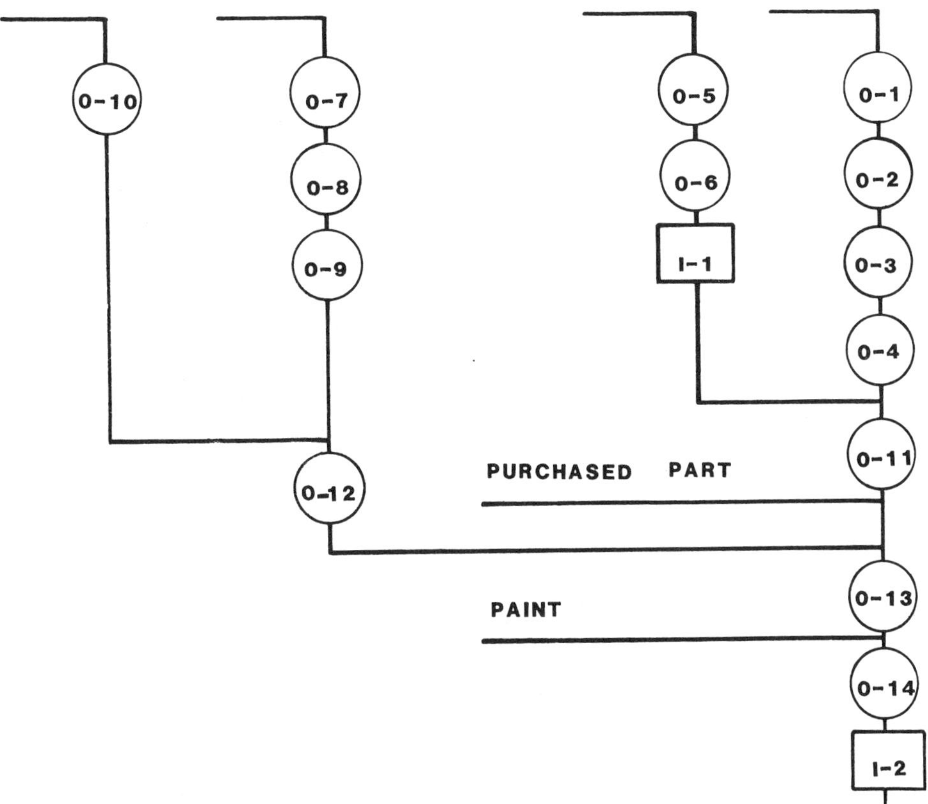

Fig. 6.11 An operation process chart.

analysis could be expanded to cover more than two items. However, if the steps in each pick are the same, it's not necessary to show each item separately. The analyst can note that certain operations are repeated a number of times. If the order picking calls for a variable number of items per order, a frequency distribution of the number of items can be used.

6.9.3 Questioning the Process

Each designer and analyst must develop his or her own methods of using the operation process chart for finding improvements and designing a new process or system. As mentioned earlier, several check- and question lists for organizing this function rationally and efficiently are available. Many analysts have found the brief list of four questions useful as clues to process improvements.

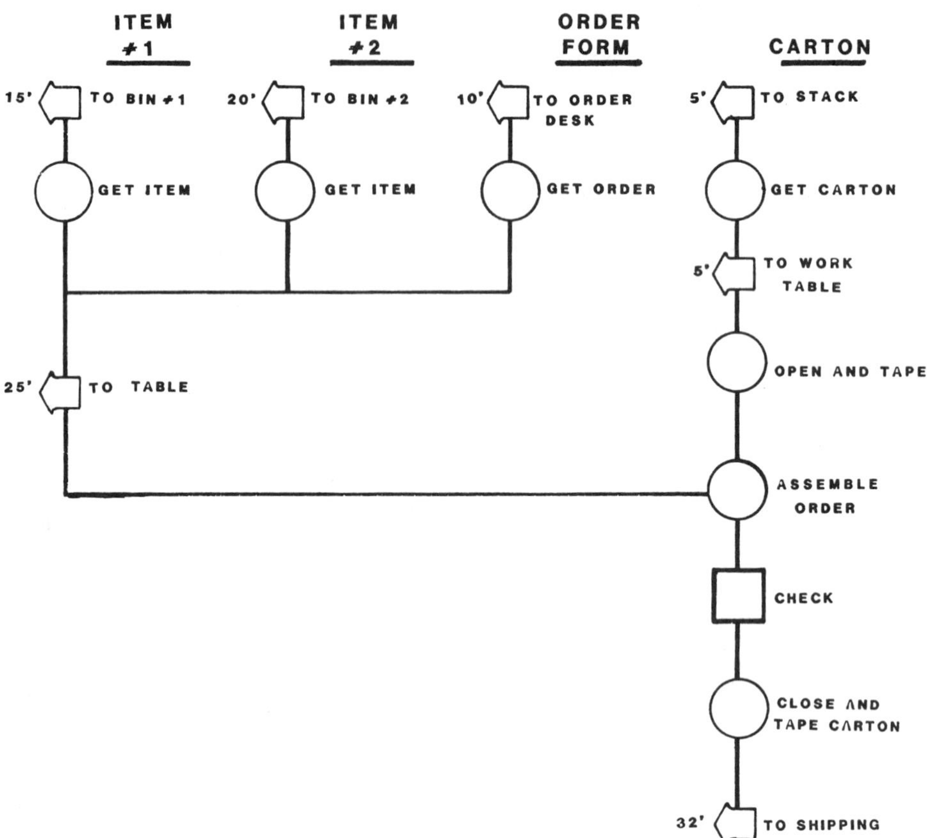

Fig. 6.12 Operation process chart of order picking.

What Can Be Eliminated?

The biggest cost reduction comes from elimination. If a process can be omitted, its entire cost is saved. Many processes and operations are continued long after their purpose and usefulness have ceased. However, unless someone reviews the process periodically for improvements, unneeded functions will remain. Perhaps only part of the process can be eliminated. If so, cost will be reduced and money saved.

What Can Be Combined?

Combining two or more operations into one can result in significant savings. A product movement may be combined with automated positioning for the next operation. An order picker may get more than one item per trip.

What Can Be Improved?

Several good checklists are available. The author favors a short and simple list. If the list is too long, even though it covers more possibilities, the analyst is likely to rush through the checklist without really thinking about its application to the process being examined.

Can the Order Sequence Be Changed?

Sometimes the order in which operations are performed can be changed to improve the process and to reduce costs. An example would be changing the order of different machining operations to reduce travel, distance, and time from one station to another in a factory laid out by functional departments. The existing processing order may have been established under quite different conditions and requirements, yet no one questioned whether the order should be changed accordingly.

6.9.4 Other Checklists

Many checklists exist for questioning and improving processes and operations. Some are published; others are proprietary. The analyst should find and use one that is most effective for him.

6.9.5 Planning the Layout from the Operation Process Chart

The operation process chart shows graphically the steps required to make the product or to complete the process. This is a guide to laying out the facility. Best results are obtained when the flow of product follows the sequence of operations shown by the operation process chart. The objective should be to minimize the travel distance and consequently travel time and expense as much as possible. The flow shown by the operation process chart will usually form the basis for a good layout and material handling system.

6.10 FLOW DIAGRAM

6.10.1 Introduction

A flow diagram tracing the path of material or information through the factory or organization is a useful tool in materials handling system design. It consists of two parts, the layout of the facility and the diagram of movement paths through the facility. The diagram of the flow may be superimposed on the layout by a transparency.

Again, simplification is advisable. If there are a large number of paths, there may not be enough symbols or colors to distinguish among the paths easily. The drawing can become too cluttered to be useful. One solution to clutter may be to make separate layout flow diagrams for each product. Another is to combine products into groups with each group having a single flow line.

6.10.2 The Layout

The first step is to make a layout of the facilities and equipment. This is necessary to ensure that the proposed system will fit properly into the facility, whether new or existing. Only a small proportion of materials handling system projects go into new facilities. The majority involve installing a new or modified system in an existing facility. If the building is new, management and designer may have some leeway in altering the basic building design to improve the materials handling system before construction begins. However, the major building features of the new layout may be fixed and allow for little or no modification.

If the project involves an existing facility, it is vital to have a complete and correct layout. All too often the actual layout does not conform to the drawings in the plant engineer's office. Alterations may have been made without being shown on the drawings. Some dimensions may have been changed during construction with no documentation. The system designer should make an on-site inspection in every case. This may disclose features not shown in the drawings, such as inadequate overhead clearances and undocumented alterations.

Building and Building Features

The first layout step is a layout of the building and its permanent features, including walls, windows, doors, and columns. This should show the location of utilities and utility outlets, overhead obstructions, and other equipment that could be moved only with considerable time and expense. Some desirable building equipment modifications are not feasible for structural reasons or because of building, fire, and safety codes. An example of a building drawing showing walls and columns is shown in Figure 6.13.

Again, it is important for the designer to check the measurements shown on the drawings with those actually existing in the plant. He should also look for items not appearing on the drawings or items shown on the drawing that have been removed.

Existing Equipment

The second step is to show existing equipment on the layout. This includes production machines, storage bins and cabinets, conveyors, scales, inspection stations, supervisors' desks, and other similar items. These differ from building and building equipment in that they can be moved with expense ranging from negligible upward if the new system calls for their relocation. Modern factory design practice is to simplify relocation of utilities and to provide for easy moving of machinery and equipment. We now recognize that a layout or material handling system is not a permanent design but a dynamic system which will be changed several times as product, processes, and markets change. Figure 6.14 shows the layout of Figure 6.13 with machines and equipment included.

6.10.3 The Flow Diagram

The flow diagram shows the flow path of the material through the factory or facility. Arrows are inserted to show direction. Each part or part family line should be in different color or code. A flow diagram added to Figure 6.14 is shown in Figure 6.15.

The flow diagram can take one of the following forms:

1. A separate diagram, as shown on the layout
2. A flow diagram drawn directly on the layout
3. A diagram on a transparent plastic

The transparency may be superimposed on the layout or removed if desired. This procedure is good for comparing several alternative flow patterns for the same layout.

Fig. 6.13 Facilities layout without equipment.

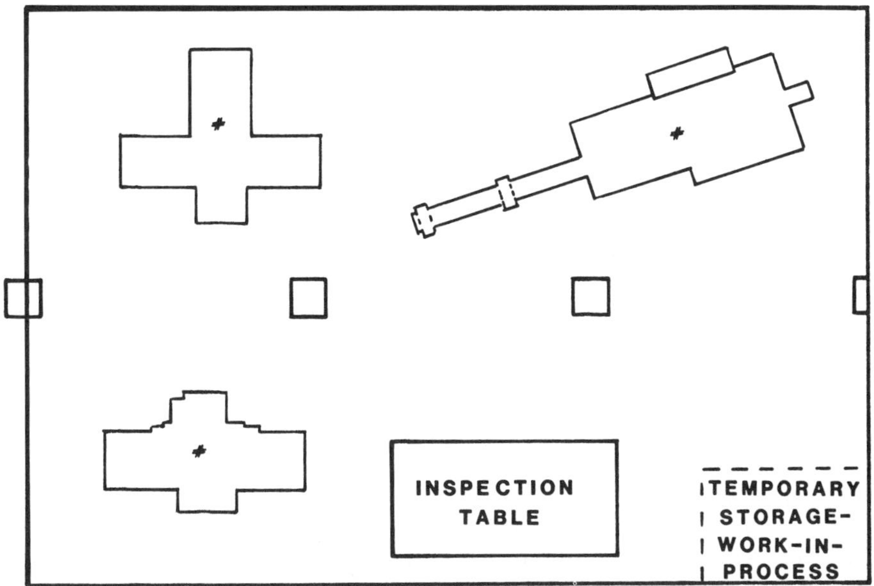

Fig. 6.14 Building layout with equipment, #—machine tool.

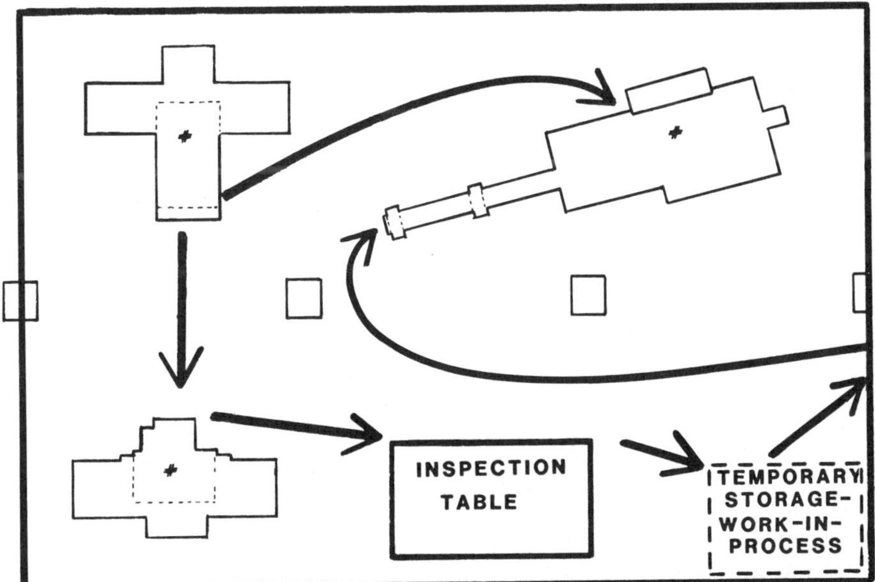

Fig. 6.15 Flow diagram, #—machine tool.

6.11 MAN AND MACHINE CHART

6.11.1 Introduction

A man-machine chart analyzes the work patterns and times for an interrelated man-machine system. It is a combination of flow process charts for both system elements, the man and the machine, to show the time interrelationships within the system. It matches the element patterns of each component to show what each is doing at any time during the cycle. For example, it shows what the worker is doing while a machine is operating automatically.

6.11.2 Purpose of the Man-Machine Chart

The purpose of man-machine chart analysis is to help the analyst determine one or more of the following:

1. The optimum combination of men and machines to achieve best economic results
2. To optimize machine utilization
3. To optimize labor utilization
4. To determine ways to improve system productivity

It's not possible to achieve all the above objectives simultaneously. The analyst must use economic data, production requirements, and other information to reach the best solution for a given situation. For example, if the major objective is to achieve production through more machine utilization, it may not be possible to optimize labor utilization by eliminating idle work time.

6.11.3 Description

A man-machine chart consists of two parts. The first is the heading with information on the job, the analyst, date of the study, and other similar data. The second is the chart of the system operations. There is a column for each man and each machine. Each element is described. Element times are entered and are shown on a sequential time scale. The times are matched to show similarity of functions.

6.11.4 Example

Figure 6.16 shows a man-machine chart for order picking from a two-bank miniload automatic storage and retrieval system. Time scales are shown for the operator and for each storage bank. Operating times are cross-hatched; idle times are plain. In the original installation, the operator must wait until each carrier is returned to its original position before calling the next carrier.

PAGE —— OF ——

MAN AND MACHINE CHART

OPERATION Order Picking from two-bank AS/RS
OPERATION NO. 234
PART NO. NA
DESCRIPTION See SOR-234
MACHINE NO. AS/RS
OPERATOR Various
PREPARED BY R.M.E.
ORIGINAL OR REVISED Original
DATE 3-6-86

OPERATION	OPERATOR	LEFT BANK	RIGHT BANK
Call container from left bank			
Container moves to pick position			
Pick items			
Dispatch container			
Call container from right bank. One from left returns			
Container from right moves to pick position			
Pick items			
Dispatch container			
Call container from left bank. One from right returns			
Container moves to pick position			
Pick items			
Dispatch container			
Call container from right bank. One from left returns			
Container moves to pick position			
Pick items			
Dispatch container			
Call container from left bank. One from right returns			
Container moves to pick position			
Pick items			
Dispatch container			
Call container from right bank. One from left returns			
Container moves to pick position			
Pick items			
Container from right returns to position			

Fig. 6.16 Man-machine chart of order picking from two-bank automated storage and retrieval system (AS/RS)—original.

PAGE ___ OF ___

MAN AND MACHINE CHART

OPERATION Order Picking from two bank AS/RS

OPERATION NO. 234

DESCRIPTION See SOR-234

PART NO. NA

OPERATOR Various

MACHINE NO. AS/RS

ORIGINAL OR REVISED Revised

PREPARED BY R.M.E.

DATE 3-16-86

DESCRIPTION	OPERATOR	LEFTBANK	RIGHT BANK

Call carrier from left bank (O)

Call carrier from right bank (O)
Carrier moves to pick position (L)
Pick items (L)
Carrier moves to pick position (R)
Dispatch carrier to start (O-L)
Call carrier from left bank (O)
Carrier returns to start (L)
Pick items (O-R)
Carrier moves to pick position (L)
Dispatch carrier to start (O-R)
Call carrier from left bank (O)
Carrier returns to start (L)
Pick items (O-R)
Carrier moves to pick position (L)
Dispatch carrier to start (O-R)

Carrier from right (O)
Carrier returns to start (R)
Pick items (O-L)
Carrier moves to pick position (R)
Dispatch carrier to start (O-L)
Pick items (O-R)
Carrier returns to start (L)
Dispatch carrier to start (O-R)

Carrier returns to start

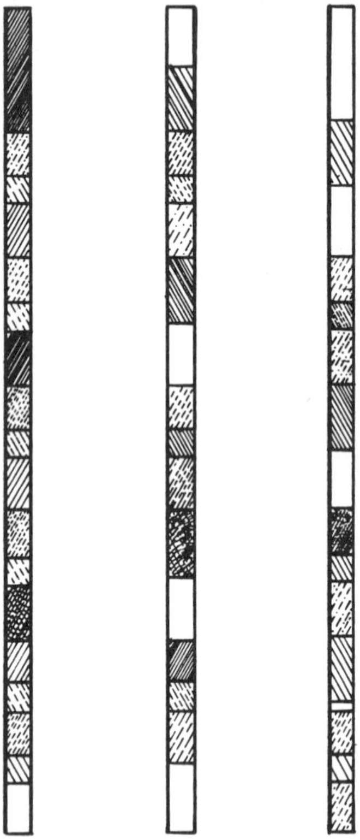

Fig. 6.17 Man-machine chart of order picking from two-bank automated storage and retrieval system (AS/RS)—revised.

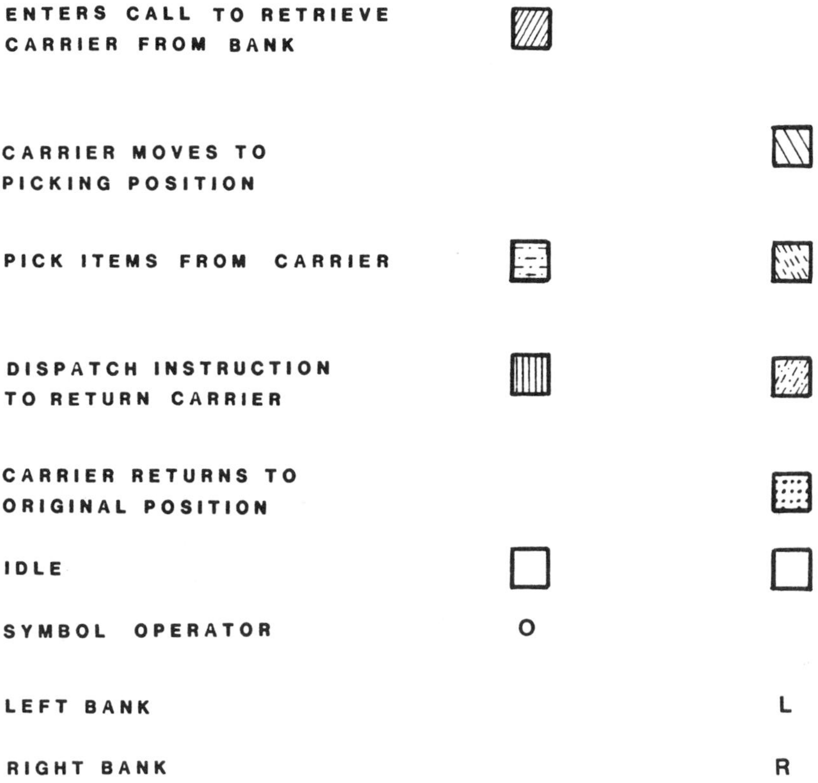

ENTERS CALL TO RETRIEVE CARRIER FROM BANK

CARRIER MOVES TO PICKING POSITION

PICK ITEMS FROM CARRIER

DISPATCH INSTRUCTION TO RETURN CARRIER

CARRIER RETURNS TO ORIGINAL POSITION

IDLE

SYMBOL OPERATOR	O	
LEFT BANK		L
RIGHT BANK		R

Fig. 6.18 Man-machine charts—key.

Figure 6.17 shows an improved system in which the operator calls the next carrier before starting work on the one in front of him. The called carrier then moves toward the pick position while the operator is working from the previous carrier. The chart shows how the productivity of the system is improved and the labor utilization increased. Figure 6.18 depicts the symbols used in the man-machine chart.

6.12 ROUTE SHEET

6.12.1 Introduction

The route sheet lists the operations and processing required to manufacture a part. They are listed in the most efficient sequence. Machines, equipment, and tooling are listed for each operation. A

ROUTE SHEET

DATE______

PART NAME ____________________ PART NO. ____________

RAW MATERIAL ______________ DRAWING NO. _____ DATE ___

REMARKS ______________________________________

PREPARED BY ________________ APPROVED BY ___________

OPN. NO.	OPERATION DESCRIPTION	MACHINE	TOOLING	STAN-DARD TIME (hrs per piece)

Fig. 6.19 Sample route sheet form.

route sheet may show times; some do not. The two time components, setup and per piece, may be shown separately. Sometimes they are combined, particularly in industries that use a standard lot size or have continuous runs without periodic setups.

The route sheet doesn't show the detailed motions and steps for each operation. This is the function of the operation sheet. Furthermore, deviations from the best sequence shown on the route sheet may occur due to production scheduling and machine availability problems.

In a modern integrated computer-aided manufacturing system, the route sheet is computerized and stored in memory. It can be retrieved and used for production planning and scheduling, procurement, personnel planning, and other purposes. An even newer development is computer development of the route sheet from computer-aided design data. This technology is not currently widely available.

6.12.2 Format

Figure 6.19 depicts a typical route sheet form. It's simpler than that used by some companies. However, it does provide the essential data on the operations and processes needed to make the part together with tooling and equipment information and the standard times per part.

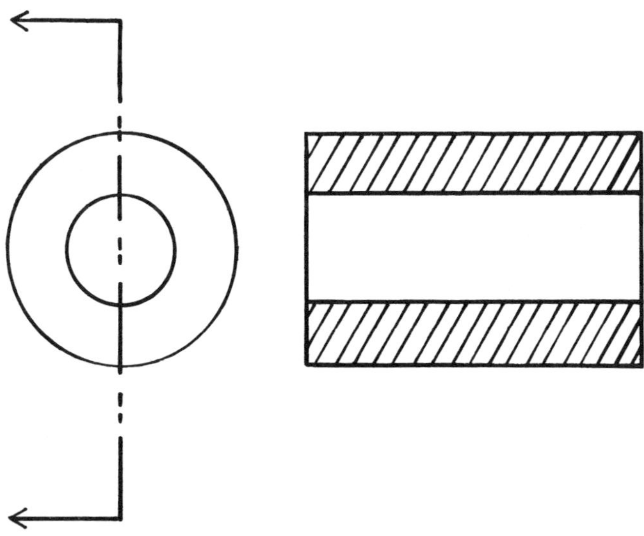

Fig. 6.20 An oil-impregnated bronze bearing.

ROUTE SHEET

DATE March 3, 1986

PART NAME Bearing, self-lubricating **PART NO.** A731

RAW MATERIAL Bronze **DWG.NO.** A731-D **DATE** Jan. 8, 1985

REMARKS Impregnated with lubricating oil

PREPARED BY D.V.R. **APPROVED BY** R.M.E.

OPN. NO.	OPERATION DESCRIPTION	MACHINE	TOOLING	STD. TIME (hrs./piece)
1	Mix powders	Mixer	None	.002
2	Compact	Compacter Press	Special Dies	.010
3	Sinter	Sintering Oven	None	.005
4	Sizing and coining	Press	Special Die	.020
5	Impregnate with lubricant	Heated Tank	None	.006
6	Inspect	-	Gages	.012
7	Wrap and package each	-	-	.008
8	Pack in shipping cartons	-	-	.007

Fig. 6.21 Route sheet for an oil-impregnated bronze bearing.

6.12.3 Example

Figure 6.20 is a simplified drawing of an oil-impregnated bronze bearing manufactured by the powder metallurgy process. After the bearing is shaped from bronze powder by compacting and sintering, it is sized and then immersed in a heated oil bath. The oil enters the pores of the bearing and provides continuous lubrication without the need for adding lubricating oil periodically. Figure 6.21 is a route sheet of the steps required to make the bearing.

6.13 OUTPUT OF THE ANALYSIS

The output of the analysis is a set of data, calculations, and information that serves as the basis for a recommended materials handling system design. However, management is usually interested only in the recommendations and the major factors involved. It seldom cares about or looks into the detailed calculations and the reasoning that went into the final recommendations. The results of the analysis are useful primarily to the designer and the checker.

6.13.1 Report

The engineering report on the materials handling system design is the channel of communication from the designer to the management. The format is usually specified by the company or by the manager decision maker. Within the specified format, the designer has considerable opportunity for effective communication with the decision maker and with others involved. Since report writing is beyond the scope of this book, the reader is referred to one or more of the many good works on the subject. Another source of help in many companies is an editorial department, which can furnish suggestions and editing.

6.13.2 Recommendations

A key element of the report is the designer's recommendations to management. The manager wants more than an engineering analysis of the problems. He wants a recommendation for action. He may not follow the recommendations, but he still wants them. The manager may have information not divulged to the designer, or he may evaluate key parameters differently with a different outcome.

The recommendations should be detailed enough to form the basis for action. The designer should consult with appropriate officials and managers during the design process. This should ensure that his recommendations are consistent with managerial thinking and company policy. If so, the probability of their being accepted is high. Some managers will want to approve decisions at several stages along the way. Others may want to see only the final product. The designer must be guided by his perceptions and knowledge of the manager's wishes and personal style.

6.13.3 Revisions

Few materials handling design projects proceed smoothly from one stage to the next. Usually, previously completed work must be revised in light of changing requirements, new technologies, market shifts, or other factors. This is normal. The designer must accept and cope with them as a part of industrial life.

6.14 FREQUENCY DISTRIBUTION

6.14.1 Definition and Description

A frequency distribution is a tabulation of the number or proportion of times various events or measurements occur. For example, a frequency distribution of the height of male workers would show the number or percent of workers 6 ft tall, 6 ft and 1 in., etc. The frequency distribution is useful in analysis and design of a materials handling system in which a parameter does not have a constant value. An example is a distribution of the number of line items per customer order. Most will be for a small number of items, yet there will be a few with several pages of line items. If the order is a unit in designing the system, the variation in number of line items per order will enter the design calculations.

Another use of a frequency distribution is in simulation of the materials handling system. Many features follow a frequency distribution. A common example is the time distribution of calls for supplies or tools. Their number per hour may be small in early morning, rise to a peak during the day, and fall off toward quitting time. In addition to those mentioned above, there are many examples of frequency distributions that are useful in materials handling system analysis and design.

6.14.2 Gathering Data

The data for a frequency distribution come from a variety of sources. Some may be obtainable from company records. For example, a sampling study of orders will produce the information for a distribution of line items per order. In this case, it's not worthwhile to make a study of all orders. The added time and expense will not be justified by the additional information gained. A sampling study will suffice. Also, management may adopt policies to encourage the customers, internal and external, to submit fewer orders with more line items per order. Naturally, this will change the frequency distribution. A common example is the handling and shipping charge on each order under $100. This encourages the customer to accumulate his purchases until the total is over $100. This will not eliminate small orders, but should reduce their number.

Another method is the work sampling study. In this method a series of random observations are taken of the status of the activity at the instant of the observation. A frequency distribution of the number of observations of each activity category will produce a distribution estimating the percent of time spent in each activity. A common example is a study of whether a system is operating or idle and the apparent cause of each observation of system idleness.

Another technique is the one-day study in which an observer watches an activity continuously for a day. He or she records the time required for each operation or delay during the day. The times for each category can then be cumulated to obtain the frequency distribution. This method is more expensive and less popular than the work study sampling method.

6.15 FROM-TO CHART

6.15.1 Definition and Description

A from-to chart analyzes materials handling movements within the factory according to source and destination. It shows the amount of material moved from each source to each destination within the system. This tabulation helps the designer locate the various functions efficiently and plan the material handling system.

The from-to chart is useful where material may take different pathways through the plant. An example is a machine shop, in which the sequence of machining operations on each product determines the movement path. It is not useful if every item follows an identical fixed path through the plant. An example of this is a bulk chemical factory making only one product in a fixed sequence of moves through specialized equipment. It is possible that a factory may have both single-path products and variable-path products in the same facility. An example is an automotive accessory factory in which the OEM product moves along a fixed path while the aftermarket (spare-parts) product may take variable routes.

6.15.2 Sample Form

The form for a from-to chart is shown in Figure 6.22. A computer can be used to prepare a from-to tabulation. Not only does this greatly reduce the clerical time, it simplifies testing alternative move

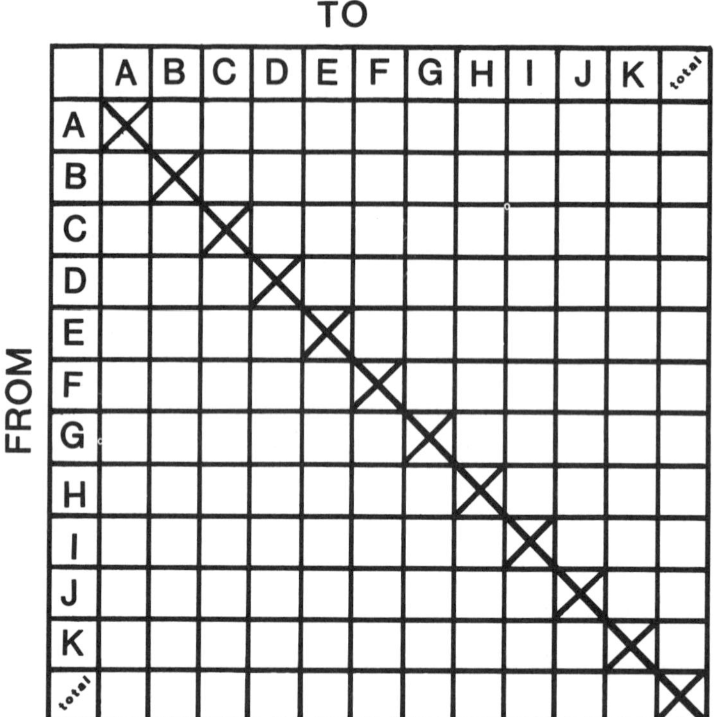

Fig. 6.22 From-to chart.

patterns. The analyst can alter the location of functions and even the order of moves. This tests the impact of proposed changes in move patterns without a physical test or extensive clerical calculations.

6.15.3 Standard Move Unit

The from-to chart must have a uniform unit of movement. If the company has a single product line, the product itself can be the unit. However, most factories make several different products of various dimensions. In this case, the standard unit must be chosen. Common ones are the pallet load and the fork-lift truck load.

A further refinement is a standard unit combining product and distance. For example, the unit may be pallet-foot, which is the movement of one pallet 1 ft. This refinement enables the analyst to test the distance effects of moving sources and destinations within the factory. Modifying the data takes time, but the potential for improvements makes the effort worthwhile.

6.15.4 Data

Seldom are numerical data on moves and distances obtainable directly from company records. A special study for the materials handling project is necessary. The objective is to get useful data of sufficient accuracy at a minimum cost in time, effort, money, and disruption. Another factor is the pressure to get the materials handling design project finished, installed, and operating as soon as possible. This pressure precludes spending much time on data gathering.

Present Situation

If the materials handling system design product is a modification or improvement of an existing system, the place to start gathering information and data is the existing system. Most such projects build on the existing system/ few are radical departures that completely discard existing equipment and operations. It is easier to observe and gather data on an ongoing operation than to make reliable parameter estimates of proposed systems that have no current counterpart.

Predicting Future Materials Handling Traffic

A more difficult data problem is estimating future traffic flow and trends. Most materials handling projects are driven at least in part by anticipated need to handle a steadily increasing volume of materials handling flow. Predicting the amount and timing of this increase is not easy and is fraught with unforeseen risks and variables.

One source of estimates of the future is the marketing department. It should have quantitative predictions of future growth and of product changes. The analyst may have to do some investigating if market predictions are unavailable. Also, he may want to verify some of the major predictions himself. The timing of increases is important. Management may want to upgrade and expand the production and materials handling system in stages to minimize current capital investment.

There are two major types of future changes: steady growth and sudden major shifts in product. An example of the first is the growth of convenience packages foods owing to increased customer affluence and to decreased time for food preparation. A major product shift was the displacement of desk calculators and slide rules by electronic pocket calculators and digital computers.

Predicting a sudden shift in product is difficult. It usually means shutdown of the old facility and manufacturing the new product in a different facility in a different location. Obsolescence of a standard product and its replacement by a radically different one often lead to the demise of the older company or division and establishment of a new company or division for the new product. Technological forecasting and assessment is a technique for evaluating expected future developments.

7

Principles of Good Materials Handling System Design

7.1 INTRODUCTION

In this chapter, we shall present a number of principles of good materials handling system design. These are guides to appropriate action, not hard and fast rules to be applied rigorously to every situation. Rules-of-thumb are convenient and easy to apply. However, the designer must be careful to assure himself that the situation and assumptions are correct for the principle in question. Charles F. "Boss" Kettering of General Motors made a fortune for his company and for himself by astute questioning of conventional knowledge assumptions and rules-of-thumb.

7.1.1 Relation of Plant Layout to Good Materials Handling

A materials handling system does not operate in isolation from its factory. It moves material from one location to another—from one function to the next. Efficient location of those functions is essential to the design of a good materials handling system. In many companies and university courses, plant layout and materials handling are combined into one department or course.

7.2 PRINCIPLES—USES AND MISUSES

A principle as used here is a guide to action. It is not an inflexible rule to be applied in every situation. Good materials handling systems embody most or all of these principles. Careful attention to principles of good design will produce an efficient and economical materials handling system.

However, two or more principles may be in conflict or even mutually exclusive in a specific application. For example, "Minimize manual handling" may conflict with "Keep costs down." A prominent automobile manufacturer built an engine assembly plant with complete automation. Within a few months, several automated features were abandoned in favor of manual handling because of the excessively high costs of the automated features. The designer must use good judgment in applying these principles to ensure the best results.

7.3 PRINCIPLES OF GOOD MATERIALS HANDLING SYSTEMS DESIGN

The following set of principles has been developed from the author's experience in teaching and consulting. Others may have a slightly different list. The advantage of a formal list is that it calls the designer's attention to a principle or concept he may have overlooked because of the press of time or emphasis on other aspects of the system design.

7.3.1 Keep Your Objectives in Mind

When the designer started his work, he had one or more objectives to accomplish with his design. These should be kept in mind all through the project. A periodic review of the stated objectives and assumptions is a good way to keep the project headed toward the right conclusion. It will help avoid being sidetracked by extraneous or immaterial issues.

7.3.2 Keep Material Moving Forward

The material should keep moving in the direction of the final destination. Backtracking, side trips, and diversions should be avoided. The cheapest, fastest, most efficient route is the direct one from start to finish.

7.3.3 Minimize the Number of Handlings

Every time the material is handled, whether manually or automatically, additional expense and time are incurred. Good design calls for the material to be handled as few times as necessary to accomplish the system objectives. The number of handlings may be reduced by combining several operations or moves, by eliminating unnecessary moves, or by introducing new equipment which will perform more than one operation on the product simultaneously.

7.3.4 Use the Right Equipment

Choice of the right equipment for the job may mean the difference between a successful installation and a failure. What is the right equipment? It is the cheapest available equipment that will accomplish the mission. In another chapter, methods of evaluation and comparison are presented.

The designer is faced with a multitude of equipment alternatives. He must call on his experience and that of others to make a good choice. Discussion with others in the organization or with persons in other companies may provide valuable information unbiased by the vendor's sales motivation.

7.3.5 Integrate Materials Handling into Other Functions

Integrate your materials handling system into other systems in the factory. These include manufacturing, inventory and production control, inspection, quality assurance, receiving, and shipping. Failure to do this may result in inefficiencies, extra costs, and suboptimization at the expense of overall operating efficiency.

Production Control

Modern communication and computer equipment enables the system to keep track of production on a real-time basis as it moves through the plant. Data on production output can be communicated to central control where the information is immediately available and corrective action can be taken directly. The system can also be designed to track production difficulties and bottlenecks as they occur. Thus management is informed currently, not days or weeks later.

Automatic counters can be used to count production output and transmit the data. These counters are accurate, more so than a manual worker. By intergrating the system into receiving and shipping, factory management can have current reports on inventory status and transactions.

Production

Combining a production operation with materials handling will reduce materials handling costs by eliminating handling and motions. For example, a finishing operation may be done as the product moves along the conveyor system.

An automatic transfer system may be a practical way to combine production and movement. The system moves the material product into position without the need for manual handling. When the production station completes its task, the transfer system moves the product on to the next station. In some automotive installations, a part moves automatically through a series of machine and finishing operations without stopping between the start of machining and final inspection.

7.3.6 Minimize Manual Handling

Use as little manual handling as possible. Manual handling is expensive and subject to error. In addition, job descriptions and definitions may limit materials handling to a specific group of employees and result in underutilization and excessive idle time. Often, simple inexpensive equipment such as a gravity trough or roller conveyor will eliminate the need for manual handling and return the investment within a short time.

7.3.7 Design Safety into the System

Design the materials handling system to protect personnel, product, and equipment against damage and accidents. Regulations such as OSHA and fire protection prescribe the safety features that must be included. Other features require the attention and initiative of the system designer.

Materials handling accounts for a high proportion of industrial accidents in the United States. These include tripping, falling, spills, and collisions. Preventive measures might be barriers, quick reactions to accident exposure, and collision avoidance devices. Product damage can be minimized through design features such as containerization, eliminating sharp corners, and smooth deceleration.

7.3.8 Keep It Simple

Use the simplest system that gets the job done at the lowest cost. As a general rule, the simplest system will be the most satisfactory in operations, maintenance, and uptime. Each added component increases the number of ways the system can fail and consequently the number of failures.

7.3.9 Keep Utilization High

Plan for maximum utilization of equipment, facilities and labor consistent with needs for maintenance and emergencies. Idle time is expensive. The more automated the materials handling system, the higher the fixed cost which goes on no matter how much or how little the system is used. In addition, the worker must be paid his regular wage when he is standing idle because of breakdown or lack of materials.

One hundred percent uptime is not a feasible goal. There will inevitably be shutdowns for routine maintenance, emergency repairs and other unavoidable causes. The goal should be to minimize the downtime and maximize the time the system is in operation.

7.3.10 Design for Flexibility

Design the system for flexibility to meet future needs and changes. Change is the normal condition, not the exception, in American industry. Few products and processes remain the same over an extended period of time. A manufacturer must be making improvements continually if he is to stay competitive in the marketplace. The engineer must design his system so that necessary changes and improvements can be incorporated efficiently without having to build a completely new system. This requires a balance between current and future needs.

There are limits to the flexibility that can be economically designed into the system. Complete flexibility is excessively costly and usually unnecessary. Changes expected in the near future are more important than those which will occur in the distant time frame. When the return on investment is considered, investment today for use several years from now is seldom worthwhile. It is also possible that the actual future will be different from the one expected at the present moment.

7.3.11 Design Capacity for Present and Future Volume

Design the materials handling system for sufficient capacity for current needs and for those expected in the near future. Increasing the capacity of a badly overloaded system is a difficult task. Sometimes, it can be done only by completely redesigning and rebuilding the system. The designer and his management should anticipate expected volume increases and plan accordingly.

7.3.12 Use Automatic Controls

Make maximum use of automatic controls to operate the materials handling system. Functions that can use automatic controls include: part-presence verification, damage avoidance, counting, and inspection. Properly designed automatic controls will have a lower error rate than a manual worker. They will be less expensive and have higher uptime and productivity.

7.3.13 Identify Your Materials Handling Costs

To the extent possible, identify the costs of moving materials separately from other costs. Costs buried in factory or general overhead will not receive attention necessary to improve operations. On the other hand, if management knows how much each move costs, it will pay attention. Managerial attention is essential to getting the managerial involvement necessary to improve material handling.

7.3.14 Use Available Space

Make efficient use of all space available including overhead cubage and yard space. Modern one-floor industrial building practice is to build up instead of out. Costs for land, foundations, and utilities are largely independent of building height. Consequently, the cost per cubic foot of going up is much less than the cost of extending outward. Higher stacking may add to material handling expense, but seldom by a prohibitive amount. Yard storage is feasible for items not subject to weather damage.

Sometimes, a company has space that is vacant or inefficiently used. The human tendency is to fill unused space even though the items stored would be better discarded or moved to less valuable storage space. Avoid using space just because it is there; make sure the space used is essential to the material handling operation.

7.3.15 Use Gravity Flow

Use gravity as the power to move material wherever possible. Gravity is strong, reliable, dependable, and free. If it's applicable, use it.

7.3.16 Design for Efficient Maintenance

Design your system for efficient and economic maintenance. Equipment recommendations should be based at least in part on the ease of maintenance. Maintainability, a new engineering concept, a concerned with the ease, efficiency, and cost of maintaining the equipment. Good maintainability will increase the efficiency of the system, improve uptime and keep costs down.

Maintenance falls into two catagories. Preventive maintenance consists of regularly scheduled inspections and routine tasks such as lubrication and changing limited life components such as filters. The other category is emergency breakdown. Quick repair of emergency interruptions is essential to steady work flow and high utilization.

7.3.17 Combine Operations

Several materials handling operations may be combined into one. Materials handling may be combined with production operations or with shipping and receiving.

7.3.18 Eliminate

Eliminate as much materials handling as possible. Every motion, movement, and operation adds to the cost. Before looking at ways to improve an operation, the industrial engineer should ask himself, "Is it necessary?" and "Can it be eliminated?"

7.3.19 Move the Shortest Distance

Move material the shortest distance required to accomplish the desired results. Time and expense are not directly proportional to the distance moved. Pickup, acceleration from a stationary position, deceleration to a stop, and setdown are independent of distance traveled. On the other hand, once the product is in motion, the time and cost of the movement are proportional to the distance.

7.3.20 Consider Personnel Requirements

Consider what personnel skills will be required by the system and how competency will be acquired. The materials handling system designer seldom is responsible for personnel acquisition and training. However, he should consider the skill requirements needed for installation and operation of his system. It may be that certain skill catagories not presently on board will be needed. Acquisition and training of persons with these skills is a management function seldom delegated to the system designer. A tight schedule with long personnel lead time may require special attention.

Good supervision is vital to successful operation of the materials handling system. However, selecting, training, and managing the supervisor is a management function, not the responsibility of the systems engineer.

7.3.21 Optimize Energy Use

Plan for optimum output of the company's energy dollar. Until recent years, the cost of power to operate materials handling equipment was a minor, if not negligible, factor in the choice among alternatives. Today, energy is a significant cost factor in many decisions; it may become even more important in the future. This would be true in a decision between an electrically operated conveyor system and a gas-powered fork truck.

7.3.22 Standardize

Standardize product, packaging, unit loads, equipment, and policies to the extent feasible. Automated, mechanized materials handling requires standard product and unit loads. Items handled must be identical within narrow limits. A manual system can tolerate a wide diversity of product and packaging; an automated system cannot.

Standard equipment helps by simplifying maintenance and repair and by reducing spare parts inventory. If all fork-lift trucks are the same make and model, personnel perform more effectively and the logistics are faster and cheaper.

Standard policies are advisable throughout the company's activities. They promote efficiency and smooth operation in material handling.

7.3.23 Develop Technological Assessment

Technological assessment is an organized methodology for looking at potential future technical developments in the company's products, processes, and equipment to assess the probable timing and impact of those developments on the company's operations. It is, of necessity, an inexact technique, yet it can be valuable in discerning broad trends and developments important to the organization. The company should provide a method and organizational responsibility for monitoring and assessing technological developments in the company's product, processes, and equipment and for preparing recommendations for coping with expected changes.

7.3.24 Plan Ahead

Plan ahead for replacement of obsolescent or worn-out equipment, for revised production requirements, for new methods and equipment, and for other future events that will influence the material handling system. Although these relate in part to technological assessment, the need for planning ahead stems from other causes as well. Physical deterioration of equipment is inevitable. A policy and plan for rebuilding and replacement should be developed. Changes in company goals and objectives may require changes in the materials handling system. New products and processes will bring other changes.

Events and decisions in other departments of the company may affect the materials handling function. An entirely new product may require a major redesign of the materials handling system. A change in sales strategy may alter the product mix and its materials handling system. Management may move into new markets, product, or geographical areas. Some changes require substantial lead time for design and procurement. Someone, engineering supervisor, general manager, or central staff, should keep abreast of changes that will affect the material handling system.

7.3.25 Use the Systems Approach

Use the systems approach in designing and implementing material handling installations. The systems approach calls for considering all the operations of the organization as a single entity rather than treating just one function at a time. The latter approach leads to suboptimization or maximizing the effectiveness of one department or function at the expense of loss of effectiveness in others. This may lead to a reduction in the overall efficiency of the organization.

7.3.26 Keep Investment Down

A key objective of management is to minimize the capital investment in equipment such as a materials handling system consistent with getting the job done. The designer must look at each item of capital investment to determine whether it is needed or whether it is the least expensive investment necessary.

This may be related to the methods of economical evaluation of alternatives discussed earlier. However, it is also related to the amount of money the company has available for capital investment and to the number of important competing demands for the company's scarce capital resources.

7.3.27 Keep Operating Costs Down

A key factor in profitable operations is keeping the day-to-day operating costs down. The system must be designed to keep these costs at a minimum consistent with available resources and with other principles of good design. The system design isn't much good unless it can control the operating costs and keep them down.

8

Designing a Materials Handling System

8.1 INTRODUCTION

Design is the next step toward an operating materials handling system that will accomplish its objectives. From a multitude of possible combinations of elements, the designer selects one set that will do the job most effectively and economically.

The design process is not a well-defined algorithm which the designer can follow through to a successful outcome. It is both a procedure and an art based on empirical experience, engineering knowledge, and ingenuity. There are some general principles and guidelines which will help the designer achieve his goal with a minimum of wasted effort and error.

One major difference between designing a materials handling system and designing a product for manufacture is the use of standard components. Most materials handling systems consist of combinations of standard components designed and made by a manufacturer of material handling equipment or systems. It is seldom economical or feasible for a materials handling system user to design the individual components. Even if the installation has unique features requiring custom-designed components, it is better to turn this task over to a company with specialized design and manufacturing capabilities.

8.2 STATUS SO FAR

8.2.1 Problem Statement, Objectives, and Criteria

The first step is to identify the problem that the materials handling system is to solve. The problem may arise from a critical incident, from management perception that improvement is needed, or from competitive pressures.

The objectives of the materials handling system are the end results that the system is expected to accomplish. A typical objective is cost reduction. Others may be better customer service, greater space productivity, or decreased accidents and damage. The criteria are those measures of the extent to which the systems design is expected to meet the objectives.

Without a clear, accurate statement of the problem and objectives, a good materials handling system for a particular application cannot be designed. Time and effort expended at this stage will save countless hours and substantial money at the later stages of the project.

8.2.2 Data

In the data-gathering phase, the designer and his support group collect information and data necessary to understand the problem and to design the new system. The type, amount, and detail of the data will be determined by the needs of the project. Too much data is expensive; too little may lead to mistakes or costly backtracking.

The data should include information on the products to be handled with specifications of pertinent characteristics, such as weight, size, shape, material, and requirements for avoiding accidents and damage. Data are needed on the space available and the building characteristics, such as utilities, columns, obstructions, and openings. Flow requirements, interfaces with other systems, and operating characteristics must be detailed. Again, the type and amount of data must be tailored to the requirements of the project.

8.2.3 Constraints

Another important determination prior to design is the set of constraints imposed on the project. Some are laid down by management, such as a limit on the capital investment or the on-stream deadline. Others are imposed by the engineering and physical principles governing the system. An example of this type is a limit on conveyor speed due to the weight and momentum of the moving product.

8.2.4 Analysis

The last preliminary step is analysis of the problem and related information gathered so far. The designer reviews what has been learned and what there is to work with. A major result of this analysis is limiting the number of alternatives to be investigated. Time and money don't permit looking at all possible options. Careful selection of the most promising choices to investigate is the key to efficient use of engineering resources and to a successful design outcome. Another outcome of this activity is determining what, if any, additional data must be collected and what, if any, changes must be made in the problem statement, objectives, and constraints.

8.3 STAGES IN DESIGN

Every system design goes through several stages from start to completion. Each stage should be completed before the next one is started. One study (1) showed that trying to overlap stages was not feasible and usually cost more in time, energy, delays, and dollars than sticking to the practice of completing one stage before starting the next.

Each stage may require managerial or supervisory approval before the next one is started. Policy varies in different organizations. Some require detailed review and close supervision at frequent intervals throughout the design process. Others are more liberal and require approval only at important decision points, such as the design concept, the capital investment proposal, and the final design.

Revision may occur in the design process. The designer may have to revert to an earlier stage and redo some previous work. This may occur due to external causes including new technological developments, change in the market, or the availability of credit. Causes internal to the design process may necessitate revisions. A design concept may prove technically infeasible after detailed analysis, or some data may prove to be erroneous or of questionable applicability.

8.3.1 Preparation

The first stage in design is preparation. Much of this is covered in Section 8.2. In addition, the designer should carefully review the resources available to do the job, the time schedule of the project, the organization constraints, and other factors that will bear directly or indirectly on the design process.

8.3.2 Design Concept Alternatives

The next step is to select the design concept alternatives to be considered. Is the system to be based on industrial fork-lift trucks, hand trucks, an overhead conveyor system, or a moving in-floor assembly line? Some principles of developing and selecting design concepts for materials handling systems are discussed in Section 8.5.

This is a stage vital to a successful project, yet one which does not require a lot of time. It is the quality of the work, not the hours spent, that determines the outcome.

8.3.3 Select the Best Alternative

After the design alternatives have been narrowed to a reasonable number, the next step is to select the most promising one as the basis for the system design. The optimum maximum number of alternatives is three to five. Too many alternatives delays the process and uses time and energy better spent otherwise. Too few alternatives risk missing the best one. This is particularly true when the manager or designer selects one alternative without adequate examination and becomes emotionally attached to that alternative.

8.3.4 Block Design

The next stage is blocking out the main features of the system. The selected design concept is broken down into major system components. For example, the designer can select the approximate length of each of several conveyors and the size of motors required without specifying the exact details.

The block design is sketched out on the layout to see whether the blocks fit the assigned area, whether there are any interferences, and whether the chosen components are appropriate individually and as a system.

8.3.5 Detailed Design

The next stage is the detailed design of the system. This includes dimensioning and specifying the components to be used. The end result is a system design ready to be implemented through purchase of parts and installation of equipment, components, and materials.

Normal engineering procedure calls for detailed checking of the final design and approval of engineering supervisors and management.

8.3.6 Implementation and Follow-Up

The designer soon learns that he is not through when the final design is approved. He must be on advisory standby during the implementation and run in stages. Changes will inevitably have to be made during installation and early operation. Specified materials may become unavailable or unacceptably delayed. Building data may prove erroneous or incomplete. A post may not be in the location shown on the drawings. The designer will be called upon to help with drawing modifications and changes.

Other changes may be required during early operations. Product damage may be higher than anticipated and require changes. Components may need adjustment or modification. Again, the designer will be called upon to help the system run smoothly and efficiently.

8.3.7 Revising the Design

Rarely is a final design the same as the original one. Revisions must be made many times as the work progresses. Causes of change include new and better data, unavailability of originally specified components, market changes, new products, or different financial limitations. At some point dictated by the schedule, the systems design must be finalized so procurement and installation can begin.

However, changes will keep coming after the design is finalized. Changes must be documented and coordinated with previous work. Modifications are best made before the design is finalized and the next phases started. Early changes are much easier and cheaper than those made later in the system life-cycle. However, changes should be made as soon as the need is perceived. Changes after installation are even more expensive in money, delayed schedules, and lost production.

A formal design change procedure is essential. Obsolete drawings and specifications must be removed from use and destroyed. Each change must be documented in an orderly fashion. Otherwise, procurement and plant work may be carried forward on the basis of obsolete, incorrect information. Correction of such mistakes is extremely costly.

8.3.8 Consultations with Other Departments

All through the design process, the project director and his associates should keep in personal contact with the individuals and departments affected by the materials handling system design. Among the reasons for this are:

1. Final acceptance and successful implementation of the new materials handling system are highly dependent on the attitudes of those operating the system and of those whose work depends on it.
2. Those affected may have special requirements that need to be incorporated. Conversely, they may know pitfalls that must be avoided.
3. Others may have some very good suggestions concerning the system.
4. Some persons may be in a position to influence management's decision on part or all of the system. It is a discouraging waste of time to have a proposed design turned down or materially modified because of objections of operating personnel, particularly if these objections could have been avoided by earlier discussions.

Discussion of the departments and persons involved is included in Section 8.15, "Interfacing Between the Design Team and Other Groups."

8.4 PRELIMINARY DESIGN

8.4.1 Selecting the Design Concept

The first step from problem formation and analysis phases to a final design ready for implementation is the selection of a design concept. According to the dictionary, a concept is "an abstract idea generalized from particular circumstances" (2, p. 272). Thus, a design concept is an abstract or general idea for a design which was derived from the particular circumstances and information previously gathered. For example, the system designer may choose to base his system on fork-lift trucks without settling on exact specifications for the trucks.

At this stage, the designer may consider several different concept options. This is the stage at which various options should be examined. Enough information has been gathered to make a distinction between promising alternatives worth further study and those which can be discarded as unsuitable without further work. If detailed study and design is done on too many options, much time and money may be spent on unsuitable alternatives with little useful result.

8.4.2 Modeling

If the designer can develop a model of the proposed system, he uses it to evaluate the system performance under a variety of conditions. A model is a representation of the system. A mathematical model which incorporates the flow rates will serve to verify the operational capacity of the system.

8.4.3 Sensitivity

A sensitivity analysis examines the variation in system behavior due to a change in a parameter. If the system behavior varies widely with a small change in a parameter, it is sensitive to that parameter. If a change in the parameter produces negligible changes in system operation, it is insensitive to that parameter.

Sensitivity analysis shows what components or features of the system should be improved or strengthened to reduce its sensitivity or vulnerability to a given parameter.

8.4.4 Compatibility

Another desirable feature of a materials handling system is compatibility with other systems and subsystems within the plant. Will the equipment be the same make and model as existing equipment? This will reduce spare-parts inventories and improve equipment maintenance efficiency. Will the conveyor system interface smoothly with other conveyor systems handling the same product? Costly manual transfers between conveyors should be avoided. Will the personnel requirements be compatible with existing job structure and job descriptions? This will make staffing, operation, and labor negotiations easier.

8.4.5 Stability

Stability is the ability of the system to maintain its behavioral pattern in spite of disturbances and to return to its original operating characteristics after it has deviated owing to a disturbance. In materials handling, this means the system can withstand disturbances without interruption or deterioration of function.

One way to obtain stability is to use a design factor large enough to protect the system from trouble if disturbances occur. There are two limits to this approach. The first is financial. Even though the factor of safety is nonlinear for many components, there is a level beyond which additional protection is not worth the added cost. The second limitation lies in the characteristics of the materials, components, and system. Exceeding a certain level of protection may not be technically feasible.

8.4.6 Optimization

Optimization is attaining the maximum operating results for a given set of inputs. For some operations research applications, optimization may be mathematically obtained and proven. For most engineering applications, optimization can't be attained with provable certainty. However, the concept is useful. The materials handling system should produce the best results obtainable with the resources committed within the established constraints.

Most optimization techniques call for a fixed set of inputs and fixed constraints. This seldom applies to practical engineering problems. The inputs change through chance or deterministic causes. These changes may alter the optimum solution. It is not practical to mathematically test the system for every change that may occur.

If simulation is used to evaluate a proposed system, the outcomes will not identify a provable maximum. However, they will help the designer determine operating characteristics of his system so he can make desirable modifications.

8.4.7 Projection into the Future

The preliminary design should be based on an estimate of future trends affecting the system. For example, increasing sales may overwhelm the system unless provision is made for expansion by increasing the capacity. Product changes may substantially affect the system.

Technological changes may influence system design. For example, improvements in computers, combined with lower costs and smaller size, have revolutionized materials handling technology. Computer developments and their effect on materials handling system design are discussed in Chapter 12. A new discipline, technological forecasting and assessment, helps organizations predict future developments and their impact. The designer should be aware of present and future changes that will affect his designs.

Looking into the future is indefinite and fuzzy at best. For most installations, a 5-year-term horizon is all that is warranted by the profitable service life of the project or the clarity and accuracy of the predictions.

8.4.8 System Behavior

One important feature of an automated system is its vulnerability to component failure. If the system goes down, the whole operation may stop. The designer must look for ways to minimize damage from shutdowns through increased reliability or through redundant circuits and components. It is seldom feasible to rely on a manual backup. It would take too many people who would be idle most of the time.

8.4.9 Testing

The designer should consider how the system will be tested both at installation and during operation. Various levels are possible, from a periodic manual visual check to a complex test with sophisticated equipment.

Testing the materials handling system and its components is a vital part of the design the implementation process. This includes the following:

1. Testing components. This is usually done by the manufacturer. For many items, testing procedures and requirements are specified by the American National Standards Institute (ANSI), the industry, or the federal or state government. In other cases, the purchaser may specify testing or accept the manufacturer's test results.
2. Installation testing. Various tests may be made of the materials handling system during installation. Some may be required by building codes. The purchaser may require others to ensure that the work is done properly.
3. Final test. This is done after the system is completed and ready for use. This should be comprehensive and thorough as changes are easier to make before the system goes into operation. If the system is in use, modification may require shutting down with loss of production time at considerable expense.
4. Later testing. Some testing or operations review come after the system is in operation. This may disclose unforeseen difficulties or conditions that call for some modification.

8.5 SOURCES OF DESIGN CONCEPTS

8.5.1 Introduction

Where does a designer get ideas for design concepts? How does he know what is available? Do these ideas come from following a well-defined algorithm based on scientific experiment? Or do they come from flashes of extrasensory perception? Neither. Let's look at some sources from which designers derive design concept ideas.

8.5.2 Experience

The designer's personal experience is the source of most design concepts. The designer has observed materials handling systems in his present and previous employer's factories. He has seen others in plants he has visited. The more experience with different systems, the wider range of possibilities that have been observed. Everyone is more comfortable with systems he has designed, worked with, or observed than with others he knows only second-hand.

Experience may limit the designer's outlook. There is a tendency to stick with a design concept with which the designer has had experience. Not only is the designer more comfortable with known concepts, he runs less personal risk when he sticks with the known than when he chooses a concept with which he has had no first-hand experience. If the new design concept succeeds, the designer will get credit, but only until his next design. If the new design idea fails, his career may suffer a permanent setback. His safest course is to stick with a concept that is known and proven.

8.5.3 Design of Similar Existing Systems

A common approach to a new design is modification of a previously successful design. The designer gets the file prints of a previous similar system, makes necessary modifications, and the job is finished. Not only is design facilitated, the risk of mistakes is lessened and the design time and expense are reduced.

This may be the most efficient approach to design. This is true if the situation is similar to the previous one. Furthermore, compatibility may be enhanced, installation and operation facilitated. On the other hand, this approach may lead to overlooking superior alternatives if materials handling technology has changed since the previous system was designed and installed. Or the designer may assume similarity when in fact the old and new are sufficiently dissimilar to require different concepts for optimum results.

8.5.4 Periodicals

Another good source of design concept ideas is the current literature. Two major periodicals are devoted to materials handling, *Modern Materials Handling* (3) and *Material Handling Engineering* (4). Others, such as *Industrial Engineering* (5), have materials handling articles applying to specific industries or operational functions. Theoretical advances are reported in professional journals such as *IE Transactions, Industrial Engineering Research and Development* (6). In addition to the articles, several periodicals carry advertising of materials handling equipment and of system designers. Often a postcard for obtaining further information from an advertiser is included in the magazine.

The reader must evaluate articles and ads for appropriateness to his problem. A materials handling system may work well in one factory but be a failure in another. Furthermore, only successes are written up. Have you ever seen an article, "How We Picked the Wrong Materials Handling System"?

8.5.5 Trade Shows and Literature

A materials handling trade show is a good oppourtunity to see many different types of materials handling equipment in a short time. An observer also has the opportunity to talk with manufacturers' sales representatives and with other show attendees. The Material Handling Institute (Pittsburgh) sponsors one big materials handling forum and show every year. Almost every industry has its own trade show with exhibits of materials handling and other equipment. Most manufacturers have descriptive literature on their products and applications.

8.5.6 Other Personnel in the Organization

Good ideas and suggestions often come from persons within the organization. The plant engineer usually has many years of experience and can suggest possible solutions. Equally important, he can point out difficulties that might pose major difficulty in design or implementation.

A recent trend in American industry has been to bring the hourly workers into discussions and plans, especially on systems that affect their jobs. Sometimes an hourly production or maintenance worker will have a good idea. Today's hourly workers are better educated and informed than their predecessors of 50–100 years ago.

8.5.7 New Combinations of Existing Elements

A different or unusual combination of previously known components may produce a good design concept for the overall system. A product may be lifted by a short conveyor to height from which gravity can operate the rest of the way. A production operation may be integrated into the materials handling system that moves the material. Most inventions are not completely new; they are new combinations of existing elements.

8.5.8 New Technology

New technology may provide a new and better system design concept. The modern computer can perform many functions previously done manually or by much slower technology. The laser has brought us automatic identification, bar codes, and scanners. New finishes and electrostatic painting have brought new opportunities to integrate finishing into a moving product line.

8.6 SYSTEM COMPONENTS

At first, a system may seem a bewildering array of many components coordinated by a mysterious set of relationships that produce a desired result. One may understand a system more easily if one categorizes its components and analyzes their relationships one at a time or in small groups. Each system may be broken down into five component categories plus various subsystems, each of which has its own components.

8.6.1 Inputs

The first set of components is input, or what is put into the system. Typical inputs into a material handling sytem are equipment, product, personnel, energy, and money. These inputs may be further subcategorized into suitable classes. For example, the personnel input may be defined by labor category and number of employees in each category.

A common denominator of inputs is money. Each input represents an expenditure of resources by the organization. Ultimately each resource is furnished by expenditures from the organization's treasury.

8.6.2 Outputs

An output is a useful result coming from the system. In materials handling, the output is normally a product movement to a place where the product has more utility to the firm than it had in its previous location. This new location may be another processing station, inventory storage, or the shipping department where it will be sent on its way to the customer.

The fundamental objective of any industrial or distribution system is to produce outputs that have greater value than the cost of the inputs. If this isn't true, the system is worthless or worse. Additionally, the system should produce its output values with a minimum of inputs requiring resource expenditures.

8.6.3 Processor

The third component of a system is the processor, the mechanism that converts the inputs into useful outputs. It may be the conveyor that moves the product along the assembly line from start to finished product. It may be a fork-lift truck moving material from its beginning position to its destination. The processor may include power sources, motors, and other items necessary to the system operation.

8.6.4 Feedback

A system cannot be started and left to run without direction or control. This control requires feedback about the operation. This includes data on volume of product moved, interruptions or downtime, personnel utilized, damaged goods, and other facets of the system's operations.

Feedback may be simple or complex. For example, a simple system may require only a periodic manual observation by a production worker. On the other hand, a complex system may call for special instrumentation to record and transmit data at a pace far beyond the capabilities of any manual worker.

8.6.5 Control

The final component of a system if control. This includes the functions of receiving, processing, and analyzing data received from the feedback component. This is not enough. Control must also make decisions on the operations of the system. For example, if it receives information that there is a jam, it must shut down the system to avoid further piling up and damage. Control may communicate its information to a worker who will analyze the incoming information, make a decision, and take appropriate action. In this case, the worker becomes part of the control function.

8.7 ALTERNATIVES

An alternative is a possible course of action. In our context, it is a design concept that is a possible solution to our problem. Usually, there is more than one alternative solution to a given assignment. If there is only one possible alternative, the selection process is eliminated but the designer may be unduly restricted.

One alternative that is often overlooked even though it may be worth considering is doing nothing. The most satisfactory solution may be to stick with the present situation. Too often management attention is directed to which alternative system design should be installed while ignoring the first question, "Is a new system needed at all?" An example is "What type of warehouse should we build?" when the first question should be "What are our storage and distribution requirements?" and then "What is the best way to meet these requirements?"

8.7.1 Reasons for Presenting More Than One Alternative

There are several reasons for presenting more than one alternative for consideration and approval. Among these are:

1. Management may evaluate pertinent factors and parameters differently than the designer does. Management may weight customer service or permanent personnel positions more heavily while the designer may place stronger emphasis on capacity and maintainability.
2. Management may have knowledge not imparted to the designer. Future product plans, financial restrictions imposed since the project initiation, or an impending merger may be known to the manager but not the designer.
3. A single design concept that looks good at first may not stack up well against other alternatives after thorough analysis. Looking at several options avoids locking the system into one that may not be best.

8.7.2 Managerial Policy and Style

The decision-making manager may have written or unwritten policies on the number of alternatives and procedures for evaluating each one. He may require that at least two options be presented, considered, and discussed. He may want an extensive list of possible options with analysis of only a limited few.

Managers differ in style. Some feel that they haven't done their job unless some change is made in the plans. One railroad executive reported that his superior had a reputation of never approving a design without change. The engineer carefully inserted an unnecessary embellishment in each project for the boss to remove without harming the engineer's basic design.

8.7.3 How Many Alternatives?

One of the most troublesome questions facing the designer is "How many alternatives shall I consider?" Too many is too time consuming. Too few risks missing the best option. Search algorithms have been developed for military missions such as aircraft surveillance of enemy submarines. However, none have been developed for efficiently identifying possible alternatives and selecting the best one in an industrial design situation. How are alternatives identified and developed?

This is discussed in the next section. The basic ingredient is the experience and skill of the designer. This is infeasible to quantify and put into an algorithm.

Various rules-of-thumb exist for the number of alternatives to be considered. Some authorities favor three or four of the most promising. Others allow five or six. In addition to the time and expense of looking at many options, a large number of alternatives usually contains some which are duplications or which are obviously unsuitable.

8.7.4 Recommending the Best Alternative

The designer is expected to select and recommend the alternative which in his judgment best meets the conditions, criteria, and constraints of the problem. It is not enough to list the alternatives and expect the manager to choose. A major part of the assignment is selecting the most promising option. The manager may not agree with the designer's choice, but he does expect the designer to have a recommendation.

8.8 FEASIBILITY STUDY

8.8.1 Introduction

A feasibility study is an investigation into the extent to which a proposed design can be carried out and to which it will be economically viable. It encompasses those facets of a project which demonstrate its practicality.

The feasibility study is a basic document for management decision. It should contain data and facts that management needs to make a sound decision. The preparer of the feasibility study should anticipate the questions managers will ask and be prepared with answers.

8.8.2 Purpose

As with any project or effort, a clear understanding of the purpose of the feasibility study is essential to effectiveness. A written statement of the purpose clarifies the project in the minds of the designer and others concerned. Is the feasibility study to be the basis of selecting a design concept from several available alternatives or is it to be an almost complete design ready for approval or disapproval? Company policy and managerial style differ greatly. The designer should establish the purpose and uses of the study before he goes ahead.

8.8.3 Contents of a Feasibility Study

The contents of a feasibility study will vary depending on the situation. Management may specify what is to be included. The designer may include information he deems essential to presentation of his case. A project involving large investment will merit more detail and information than a smaller one. For very small projects, a simple memo or even a short oral discussion will suffice.

However, most feasibility studies will include at least the sections listed below. Terms and breakdowns differ among organizations, but the information remains similar even though format may change.

Description of the Project

The first introduction to the report should be a narrative description of the project. It should identify the project, discuss its background and the reasons for the project, and present the major features.

Along with the narrative, the report should present the important conditions surrounding the project. The goals and objectives of the proposed system and the criteria and constraints should be listed. The report may include information on the design team, project organization, and other administrative details of the project.

Technical Feasibility

The first question to answer is "Is this project technically feasible?" If it isn't, there's no point going further. Technical feasibility includes the ability of the system to move the product, maintainability, reliability, and ability to withstand loads and forces.

Economic Feasibility

Economic feasibility is the ability of the project to earn back the investment plus a return acceptable to management. Unless the project is economically worthwhile, technical and operational feasibility is immaterial. Methods of assessing economic feasibility by engineering economy methods are discussed in Section 9.3.

Management may have criteria of economic feasibility other than return on investment. Chief of these is improvement in competitive position. This objective is not easily measurable through conventional engineering economy methods, but it is vital to the company's survival and prosperity. Management assumes that improved competitive position will be accompanied by improved profits and financial condition. This assumption is normally warranted.

Operational Feasibility

This is defined as the ability of the materials handling system to meet operational requirements such as units of product moved per hour.

Other criteria may be maintainability, as measured by the cost and simplicity of performing required maintenance, and reliability, or minimum downtime, of the system.

This parameter merges into technical feasibility. There is a distinction. Technical feasibility is the ability of the system and its components to meet the required strength and similar specifications. Operational feasibility is its ability to meet the movement requirements of the production system of which the material handling system is a part.

Other Factors

Included under this heading are characteristics and data that do not fit into one of the previous categories. Examples are downtime, delivery time, personnel requirements, training, and information on similar systems if available. This information may be requested by management or furnished by the designer as a necessary part of the study.

Alternatives

In this section are listed the alternatives considered but not recommended. Each should have a brief description and a statement of reasons why the selected option was recommended over a rejected one. Some managers may desire that this section be omitted. If so, it is left out.

Recommendations

The final section of the main report is the set of recommendations. The designer should list his recommendations with respect to the materials handling system he has selected as best meeting the organization's needs. These should cover the design, the investment, the schedule, and other major points requiring management approval. The number of recommendations should be limited to not more than five or six. Too many is confusing and involves too much detail. If advisable, several minor points may be combined into one major recommendation, with subsidiary ones being explained in detail in an appendix or supplementary section.

Appendices

An appendix is the place to include details of data, design calculation, extensive references, or other material that can be efficiently separated from the main body of the report. Examples are pertinent sections of an applicable building code, manufacturer's literature and specifications on major equipment items, and detailed data on product specifications and production rates. The appendix is also useful for information that is to be omitted from some copies distributed to certain persons and departments.

8.9 DETAILED DESIGN

After the preliminary design has been completed, the design concept selected, the feasibility study made and approved, the designer's next task is to fill in the details. The end result will be a set of designs, drawings, and specifications from which materials and equipment can be purchased and the system built and installed.

The organization may elect to use a vendor's design services for the detailed design. This may be the most efficient mode if the vendor is suppling the entire system, including purchasing or manufacturing the physical components and supplying the labor. If this arrangement is used, the purchasing company should insist on getting a set of drawings and specifications for itself. This is essential for efficient maintenance and modification of the system after it is installed.

8.9.1 Subsystems

Detailed design should proceed in hierarchical fashion. That is, the overall design concept is selected first. Next, each of the primary subsystems is defined. Then, the minor subsystems, if any, are specified. Finally, the individual components are selected. This order is essential to good design. Otherwise, a component choice may be negated by the selection of a subsystem of which it is a part.

For example, a concept using fork-life trucks has been determined to be most efficient and cost effective. The next step is to decide what type of vehicle is best, what fuel, what size, and what capacities. Next is to select the recommended make and model, especially if there are substantial differences among fork-life truck vendors. Finally come decisions on special features such as attachments for handling the system's products.

8.9.2 Components

Once the subsystems are settled, the individual parts and components that go into the system must be designed. Most materials handling system components are standard off-the-shelf items bought from specialized vendors. It would be ridiculous for an assembler of appliances to design and build his own fork trucks. The same principle holds for almost every materials handling system component. The user doesn't have the volume, the design and manufacturing know-how, or the facilities to make materials handling system components himself.

Occasionally, an organization may need a system or subsystem that is unique and unobtainable from the vendor of systems or components. In this case, it may be advisable for the buyer to design what is needed and put the item out for quotes from specialists. Other situations may arise. The company may want to utilize its own labor to build a simple installation of racks or shelves, or it may want to avoid contracting out a design and installation that might compromise confidential data.

8.9.3 Detail Part Drawings

Conventional engineering practice calls for a separate drawing for each part that goes into the system plus subassembly and assembly drawings showing how the parts and components fit together. This is time-consuming and expensive. Today's practice is to reduce engineering and drafting time by methods that do not impair the usefulness of the information.

Ways to reduce the cost of design work are:

1. Use written descriptions and specifications in lieu of drawings.
2. Use standard items with references to that standard in lieu of a drawing.
3. Group technology in which parts are designed as a unit in a family of similar parts.
4. Rely on vendor's drawings and specifications.
5. Computer-aided design with hard-copy output where necessary.

In each case, the required output is accurate and complete information usable by those who must acquire and install the material handling system.

8.9.4 Assembly Drawings

An assembly drawing is, as its name implies, a drawing of the entire assembly and its component parts. A major purpose is to ensure that the component parts fit together properly and that all parts and fasteners are included. The designer can check clearances and discrepancies. From the assembly drawing the parts list is prepared and checked. Also, the designer can plan the parts drawings to ensure that each part has its own drawing or is otherwise properly specified. The assembly drawing serves the installation crew by showing the way the parts go together. The assembly drawing may also include notes on points that are not clear from the drawings.

8.9.5 Experimental Construction (Mockup)

In a few cases, an experimental setup (or mockup) may be necessary and useful. It is appropriate where field installation may show an interference, a lack of clearance for personnel, or other conditions requiring expensive changes in design plans and installation. An example is a materials handling system operating on several different levels. Possible interferences and clearance difficulties are difficult to visualize with two-dimensional engineering drawings but can be seen readily with a three-dimensional model.

Some companies use three-dimensional scale models for planning and displaying the materials handling layout and system. Commercial models are available at a scale of 0.25 in. to the foot. A scale model is useful as a presentation tool and a public relations exhibit.

8.9.6 Product Test Report

Seldom does the user of a materials handling system perform a test of the components prior to installation. If the product is a new, untried one made by a vendor, the purchaser may ask for a copy of the test report if such is available to a buyer. If the new component is made by the user and is new and untried, the company may run tests before it is installed. This is seldom done as few really new items are designed and built by the user of a material handling system.

This test should not be confused with the run in test made after installation and before full production operation. This is standard procedure whether the system is purchased from an outside vendor or is built by in-house personnel.

8.9.7 Analysis and Prediction

In this phase, the engineer examines the completed design and reviews the calculations. The purpose is to ensure that the system will meet all the requirements, criteria, and constraints imposed on it. He should also predict how the system will operate under normal conditions and under unusual conditions that may reasonably be expected to occur from time to time.

8.9.8 Redesign

Unfortunately, very few designs progress without alteration or redesign. There are many possible reasons for change. A specified part may become unavailable. A line supervisor may request a change to make labor operations simpler. Calculations or review may dictate a modification. It is common for a system to have a number of changes before the job is completed.

An important procedure in any engineering department is drawing control. The location of each copy of each drawing should be recorded and kept up to data at all times. When a change is made, all obsolete drawings must be withdrawn and destroyed. Otherwise, the obsolete drawing will remain in use. Work done from an obsolete drawing will be incorrect and have to be scrapped and replaced at unnecessary cost, great or small, to the company.

8.10 INTEGRATING THE SYSTEM COMPONENTS

The materials handling system designer has put together a number of different components to form his system. As a final step, he must make sure that the components are compatible and will work together. This is particularly important if the system uses a number of off-the-shelf components from different manufacturers. It is entirely possible that a pair of items that should mesh will not fit together and operate harmoniously.

One area in which integration is important is the computers, microprocessors, and controls. Some computers, software, and peripherals are not compatible. This may not be a major problem since most computers are compatible with the products of IBM, the dominant computer manufacturer in the United States.

It may be possible to integrate components through special adaptors or interfaces. Although this may be feasible, it is seldom as satisfactory as designing full integration into the system originally.

8.11 DESIGN REVIEW

A final step is check and review of the design. Checking an engineering design is a well-defined, standardized procedure in most organizations. It calls for an engineer other than the person who did the original work to go through the calculations, dimensions, and drawings to be sure there are no errors of omission or commission.

Another desirable procedure is to review the proposed design with those officials affected by the system and its operations. If coordination through the design procedure has been maintained, there should be few, if any, changes at this stage. Some standard operating procedures call for review by legal, tax, safety, or other departments having a specialized function.

Another review is the capital budgeting process. In this procedure, the economic feasibility, costs, and benefits of the project are reviewed to determine whether the proposed project will be a wise use of the organization's resources and will fit into the organization's goals and plans.

8.12 THE DESIGN REPORT

The final design report contains all the drawings, specifications, and details necessary to acquire the new materials handling system. Those in charge of installing the system should be able to work from the final design report and accompanying documents without having to refer to supplementary instructions and documents.

The format and content of the final report are normally specified by company policy. In content and organization, the final report is similar to the feasibility study. The major difference is that the final report represents completed design work whereas the feasibility study contains only the design work needed to make a sound management decision.

In this book, we shall not describe the final report in detail. The reader should refer to his company report instruction manuals or to one of the many good books available on engineering report writing.

8.13 SECURING APPROVALS

The last, but not least, important step is securing the necessary management approvals to acquire and install the new system. Without this approval, the project is dead. However, if the designer has kept in touch with management and its desires during the preceding phases, approval should be forthcoming without difficulty.

At the beginning of the project, the designer should make certain of the following:

1. Who must approve the project.
2. Who has veto power over part or all of the project.
3. Who has the authority to allocate the funds necessary to implement the project.

If these are clearly understood at the beginning, producing a successful design that will win approval becomes much simpler.

8.14 DESIGN ORGANIZATION

The design of a materials handling system requires the efforts of many people. The design team organization is the vehicle for combining and coordinating the work of these people to reach a successful conclusion—a good design on time at a satisfactory cost. In previous sections, we have discussed the project phases and organization. Here we shall cover the personnel organization.

8.14.1 Objectives of the Design Organization

As with any other group effort, people assigned to the materials handling system design project need to understand the objectives of the project and the organization. The objectives should be embodied in a clear, unambiguous, logical written statement,

Typical objectives are:

1. The design output expected of the group
2. The time schedule to be met
3. The objectives of the system

The design team should also understand any constraints placed on their work. Examples are:

1. Rules concerning contacts with potential and actual vendors of equipment
2. Internal protocol if important
3. Policies concerning expenditures for personnel, overtime, supplies, and models for the design project
4. Constraints on the design such as floor space available

8.14.2 Functions of the Design Organization

In addition to the design work, the project organization may have other functions assigned or implied. These might include:

1. Records of time spent, accounts to be charged, and other similar accounting and administrative reports
2. Performance evaluations of team members by the project director
3. Informal and formal reports of trends and developments to management if requested

In some companies, some housekeeping and advisory functions may be performed by other groups such as cost accounting, personnel, and timekeeping.

8.14.3 Organization Structure for Design

The organization structure of the design team will depend on several factors. A large project will call for more formally defined organization than will a small one. Personnel practices regarding transfers and temporary assignments may differ. Two major principles should govern every case: The organization should be as simple as possible to get the job done and should be definite with no gaps or overlaps in assignments, so that every member of the team understands his or her functions, responsibilities, and authority.

Typical design organizations are shown in Figures 8.1 and 8.2. The first portrays a small team doing a small project in a short time, the second a larger team needed for a larger project over a longer time span.

8.14.4 Staffing the Design Team

The materials handling system design team may be either a continuing organization or an ad hoc group for a specific project. Staffing procedures will differ between the two cases.

If the design team is a continuing one, moving from project to project as one finishes and another starts, more specialization of personnel can be introduced and developed. Since a person can be expected to do the same job for several years, care in selection and training are warranted. A company may have a skeleton organization of key materials handling design specialists supplemented by temporary assignments for overloads.

If the design team is ad hoc for a single project, management must make up the team by special assignment from existing personnel. This requires assembling, training, and scheduling personnel efficiently. This method is often the best way when personnel are available on a temporary basis and they are knowledgeable about materials handling and about the company's operations. Management must also plan for reassignment of design team personnel after project completion. Otherwise, temporary personnel will stretch out the work as long as possible or until they see another assignment starting when the current one is finished. Figure 8.3 shows how a permanent design team cadre can be supplemented by personnel on temporary assignment.

8.14.5 Managing the Design Team

A basic principle of management is unity of command. The project director should be responsible for the project in all its phases, technical and administrative. To go along with this responsibility, he should also have authority. The director will be limited in several ways, such as overtime policies, requirements to use employees presently on the company's payroll, and government regulations.

Top management may not allocate resources and authority needed for complete work. This is unfortunate, since parsimony at the design stage will result in less-than-optimum operating costs many times greater than the money saved originally.

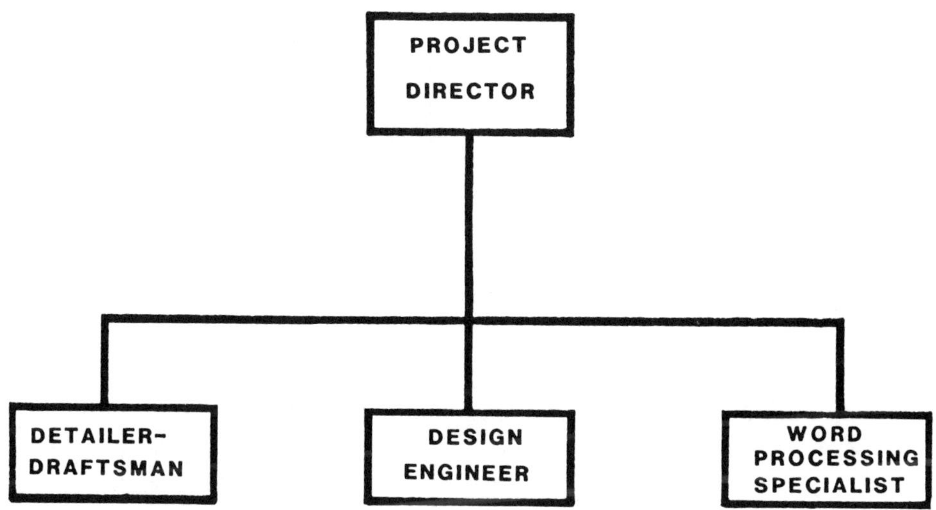

Fig. 8.1 Design team for a small materials handling system design project.

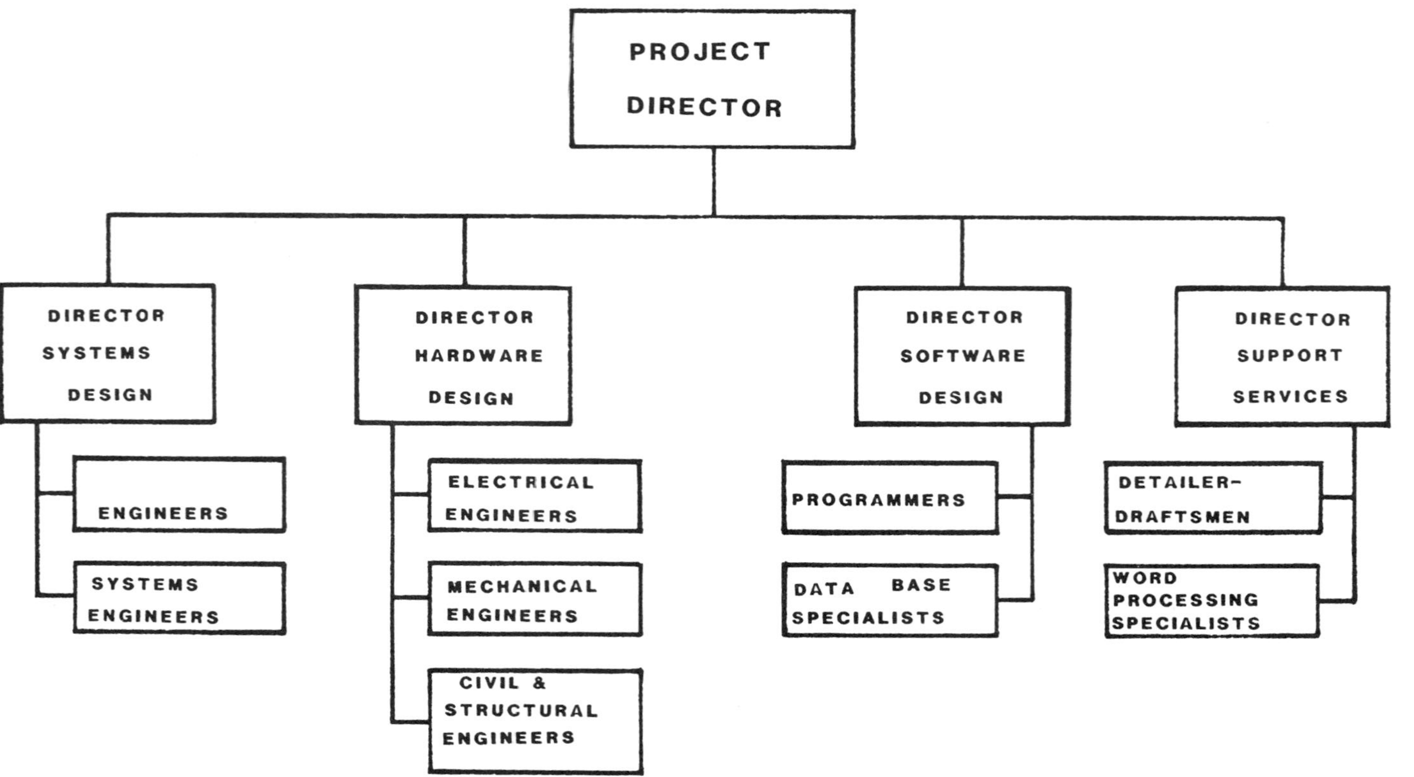

Fig. 8.2 Design team for a large materials handling system design project.

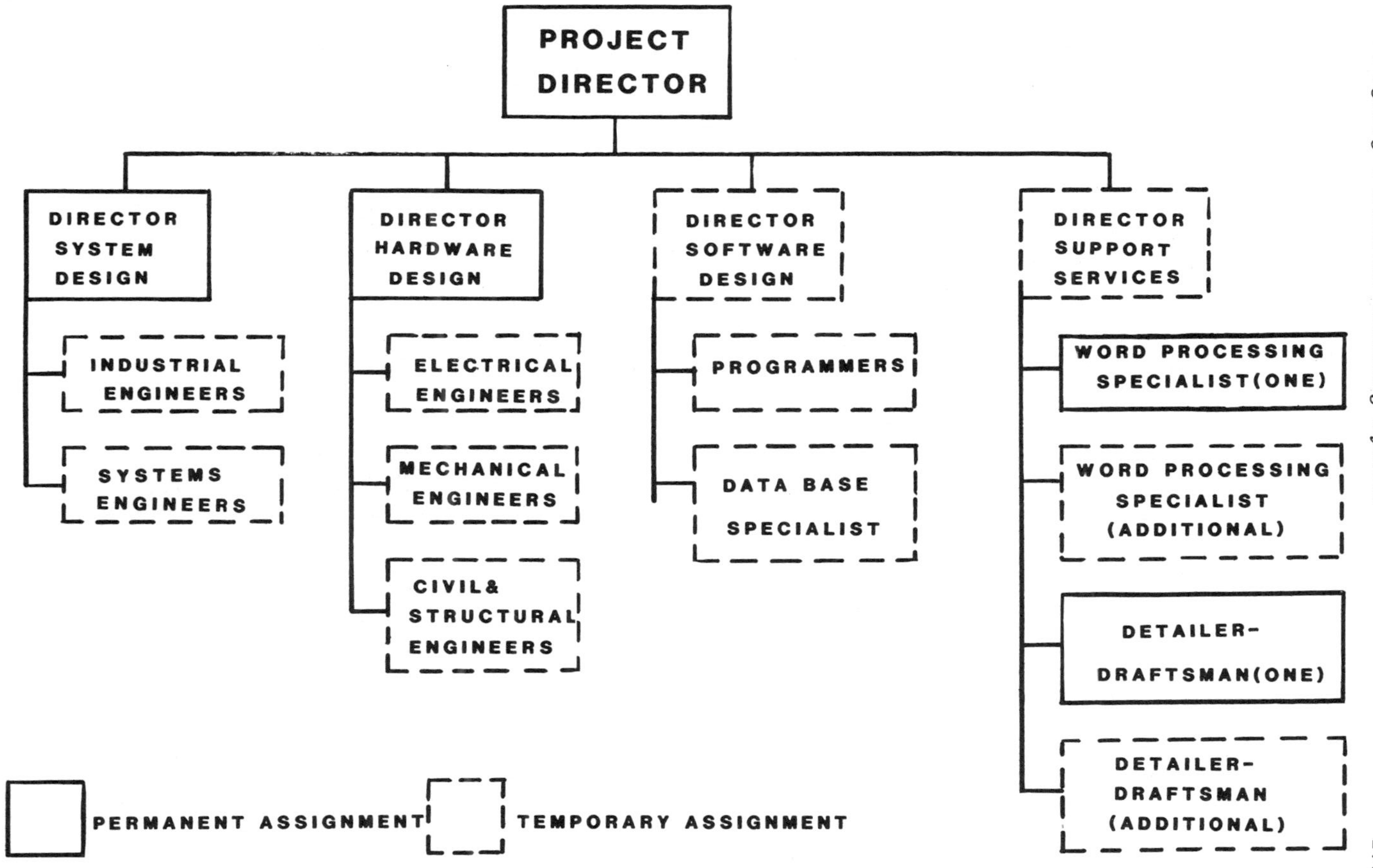

Fig. 8.3 Design team augmented by temporary assignments.

8.15 INTERFACING BETWEEN THE DESIGN TEAM AND OTHER GROUPS

8.15.1 Introduction

The materials handling system design team cannot work in seclusion from other groups and persons within and without the company. Gathering information, ideas, and comments from others will require considerable time and effort, yet this activity is essential to project success.

By interfacing, we mean the numerous contacts, formal and informal, between project design team members and others on matters concerning the project. Examples are gathering data and information, special requests and requirements from affected parties, ideas on system features, and informal approvals of proposed designs.

8.15.2 Interfacing Groups

In this section, we shall discuss interfacing with various groups within the company. These groups may be organized and designated differently from company to company. However, necessary functions are universal and are present no matter what the name or place in the organization.

Plant Engineering

Plant engineering is the department charged with the care, maintenance, equipment installation, and alteration of the company's physical plant and equipment. Another duty is verifying the performance of subcontractors and vendors of equipment installed by vendor personnel. The department is headed by the plant engineer (or facilities director) with sections by specialty such as electrical, millwright, mechanical, plumbing, and janitorial. He may have a staff of engineers, schedulers, clerks, and others to assist him.

Plant engineering is the department that must install the materials handling system with its own personnel or supervise its installation by vendor or contractor workers. Its personnel can provide information and advice on design features which will facilitate installation and operation of the system and help avoid expensive mistakes. The cooperation of each worker is important. If he is consulted and goes along with the design, he will put more effort into seeing that the installation goes well. If he is not favorably disposed toward the design team, he has many subtle ways of showing his displeasure and inhibiting the job.

Line Supervision

The line supervisor is the person who must use the system daily to meet his production goals. His requirements from the system are important—if the system doesn't produce, it is worthless. The line supervisor is a key factor in system success. If he doesn't cooperate, he can undercut the system in many subtle and often undetectable ways. He may also say "I know it wouldn't work, but nobody asked my opinion."

Quality Assurance and Inspection

Quality assurance and inspection have become increasingly important in materials handling system design. The old shop practice of moving products to a physically separate room for inspection and then back to the shop floor is all but obsolete. It was too costly and time-consuming. Today, the function is integrated into the production system so that the part moves into an automatic inspection station in the line of flow and then out to the next operation. Inspection results and production throughput can be recorded automatically by a microprocessor system. An added function of the system controller is a decision routine that will direct appropriate action if the input data meet specific action criteria.

The system designer must consult with quality assurance and inspection on the same basis he consults with the line supervisor, since these functions are now part of the production flow. Smooth product movement, proper positioning, damage avoidance, and equipment access are all important to efficient functioning of inspection.

Accounting and Cost Accounting

Why would the materials handling system design and layout be of interest to the accounting and cost accounting departments? These departments collect, analyze, and allocate costs of manufacturing. They are interested in any factory change that will affect the accounts they keep. If departmental boundaries are changed or eliminated, the accounts may also change. Furthermore, accountants will be interested in capital investment, depreciation allowances, operating costs, and other financial factors which go through their system.

Although the accountants will seldom have specific input requirements to the system design, they may help with comments and ideas from their point of view. A more practical point—an accountant may influence the stated financial outcome of a project through his allocation of overhead and costs.

Safety and Damage Control

Safety or personnel and prevention of product damag are important features of any materials handling system. The safety engineer can advise on OSHA and other safety requirements and how they can be incorporated into the new system. Often these requirements are simple and easy to meet. Furthermore, studies have shown that productivity is better if personnel are not worried about the possibility of an accidental injury.

Damage control may be combined with safety in the same department or may be a separate function. The system designer must make sure that his system eliminates, to the extent feasible, the possibility of product damage due to the materials handling function. Not only are damaged goods an expense, they interfere with the production schedule and cause delay problems with customer deliveries.

Purchasing

Standard operating procedure in most organizations requires that all contracts with outside vendors including purchase orders go through the purchasing department. This requirement calls for close cooperation between this department and the materials handling design team through all phases of the project from start to finish.

Some purchasing functions are:

1. Information on vendors on price, reliability, performance, delivery record, and other factors
2. Guidance in drawing up specifications and orders for purchased items
3. Preparing the negotiating contracts with outside firms for design and installation services
4. Negotiating specifications, price, and terms of change orders that become necessary during the course of the job

Sales and Distribution

The sales department is concerned primarily with the output of the production system of which materials handling is an integral part. It wants salable goods of high quality delivered on time. If the system can accomplish this, sales and marketing are not concerned with the design and operating details of the production system. However, the sales department can provide information on their requirements which can be helpful to the designer. Furthermore, they may have information on market changes and new products which will change the materials handling system in the near future.

Distribution is moving the product from production or inventory to the customer. It may be regarded as an extension of the production system since it continues the product from the completion of production processing on out the door to the customer. As part of the total system, distribution is concerned with the materials handling system especially with regard to packaging, damage avoidance, rates of flow, and movement to finished-goods inventory and to the customer.

Industrial Relations and Personnel

Industrial relations and personnel need to be consulted on materials handling system design features that affect the number, skills, and job descriptions of those who operate the system. Labor is the biggest component of materials handling operating cost. Usually, a new system will improve productivity by decreasing the number of workers needed. If training for new skills is needed, personnel will have to arrange for it. If new workers are to be added, personnel and line management must prepare the paperwork, get management approval for the new positions, and recruit the added workers. Personnel can also advise on facets of the system that need attention and possible modification from the job assignment point of view.

Industrial relations has the job of negotiating with the union representing the affected workers. Working conditions such as job descriptions, duties, environment, seniority, transfers, upgrading, and training are subject to collective bargaining under existing laws and regulations. This department can advise the designer on which, if any, features of the system may cause problems in union negotiations.

8.16 COMPUTER-AIDED DESIGN IN MATERIALS HANDLING

8.16.1 Definition and Description

Computer-aided design (CAD) is a technology that uses computers to design parts, components, assemblies, and systems. It was originally developed in the defense industry sector and has since spread to nonmilitary applications.

The components of a CAD system are:

1. A computer
2. A cathode ray tube or similar display
3. An input device such as a keyboard or light pen
4. A database of engineering data and formulas pertinent to the item being designed.

8.16.2 Criteria for Successful Use

CAD is not appropriate for all engineering design applications. Some criteria for successful use of CAD are:

1. An adequate database containing the information and formulas for the item being designed
2. Lower cost in comparison with other design modes such as manual design
3. Trained engineering and operating personnel
4. Software adequate to the task

8.16.3 Discussion

There are a number of software programs for materials handling and plant layout systems design. Some early ones are no longer available. Some materials handling systems design software packages are proprietary. The manufacturer of the system hardware may have a program which he can use to design a system for the customer, but not permit its use outside his company. Others are available by license or purchase. The price varies widely as does the capability of each program.

New programs are under development. They can do layout design, materials handling system design, and system simulation. The fixed-path configuration of many systems, including AS/RS, AGVS, and FMS, is much better suited to CAD and computer simulations than variable-path configurations such as one based on driver-operated fork-lift trucks.

One program to analyze the inventory parameters to determine storage requirements is based on a representative sample of the inventory. Input data include the physical measurement of the item, the activity of the item, and the size of the bin holding the part. The computer calculates the total space required and the number of containers or unit loads for each part. The program then extrapolates the sample data to estimate the total space needed and the overall space utilization. Further discussion, descriptions of specific programs, and examples appear in Chapter 19.

REFERENCES

1. R. M. Eastman, Engineering Information Release Prior to Final Design Freeze, *IEEE Transactions on Engineering Management*, Vol. EM-27, No. 2, pp. 37–42 (May 1980).

2. By permission. From Webster's Ninth New Collegiate Dictionary © 1986 by Merriam-Webster, Inc., publisher of Merriam-Webster ® Dictionaries.

3. *Modern Materials Handling*, Cahners Publishing Company, Division of Reed Holdings, Inc., Newton, MA.

4. *Material Handling Engineering*, Penton IPC, Inc., Cleveland, OH.

5. *Industrial Engineering* magazine, Atlanta, GA.
6. *IE Transactions, Industrial Engineering Research and Development*, Atlanta, GA.

9

Evaluating Alternative Materials Handling System Designs

9.1 INTRODUCTION

Evaluation is defined as "to determine or fix the value of" or "to examine and judge" (1, p. 429). Our objective is to determine the value of alternative materials handling systems so we can judge which is best for the organization. The materials handling system designer rarely has the authority to approve a proposed design for acquisition and installation. Therefore, we assume that we are evaluating the proposals to make a recommendation to management. At the same time, we want to establish the value of other systems so management can choose rationally if for some reason the recommended choice is not acceptable or feasible. The process is diagrammed in Figure 9.1.

9.2 PURPOSE OF EVALUATION

The immediate purpose of an evaluation of a materials handling system proposal is a recommendation leading to a decision. The final recommendation may be a choice among alternatives, a choice with modifications, or even a rejection of all proposed systems. The evaluation looks at the degree to which each proposal meets the objectives, constraints, and criteria originally imposed.

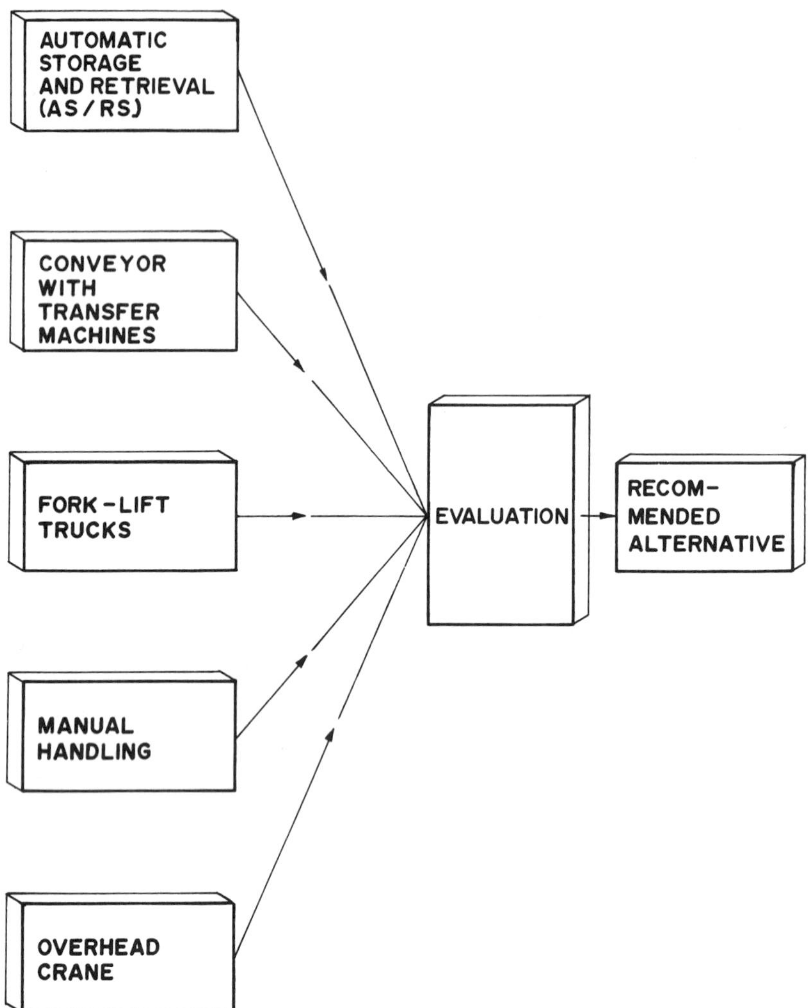

Fig. 9.1 Evaluation of alternatives.

9.3 METHODS OF ECONOMIC EVALUATION

The value of a materials handling system is expressed primarily in economic terms. Will the benefits (operating cost savings) exceed the costs? By how much? (See Figure 9.2.) Other factors enter into value; some are expressible in quantitative terms. Other factors (called intangibles) are those for which quantitative values are unobtainable or involve data for which collection is too expensive for their contribution to the evaluation.

An example of an intangible is working conditions and morale. Improvement may lead to lower turnover, improved productivity, and better quality. These are difficult, time-consuming, and costly to quantify. Furthermore, isolating improvements due to a single new materials handling system from other causes is usually infeasible. Another factor often looked upon as intangible is future change in the product or the market.

9.3.1 Payback Period

In this method, the evaluation criterion is the length of time required to recover the investment through net operating savings or net increased income. Quick payback is particularly important in times of capital shortages and high interest rates since it related to the vital factor of liquidity. It may be misleading if applied to a long-lived asset such as a building.

A number of companies use this method either by itself or as one measure among several. One conglomerate uses a combination of payback period, return on investment, and degree of risk. Another common policy is "A methods improvement must pay for itself in 1 year or less." In shoe manufacturing, the payback period for a methods change may be as short as 1 month.

Example

A materials handling system report carries the following estimates:

$$\text{Installed cost} = \$14{,}500$$

$$\text{Annual savings} = \$18{,}000$$

$$\text{Annual operating costs} = \$\ 3{,}200$$

What is the payback period?

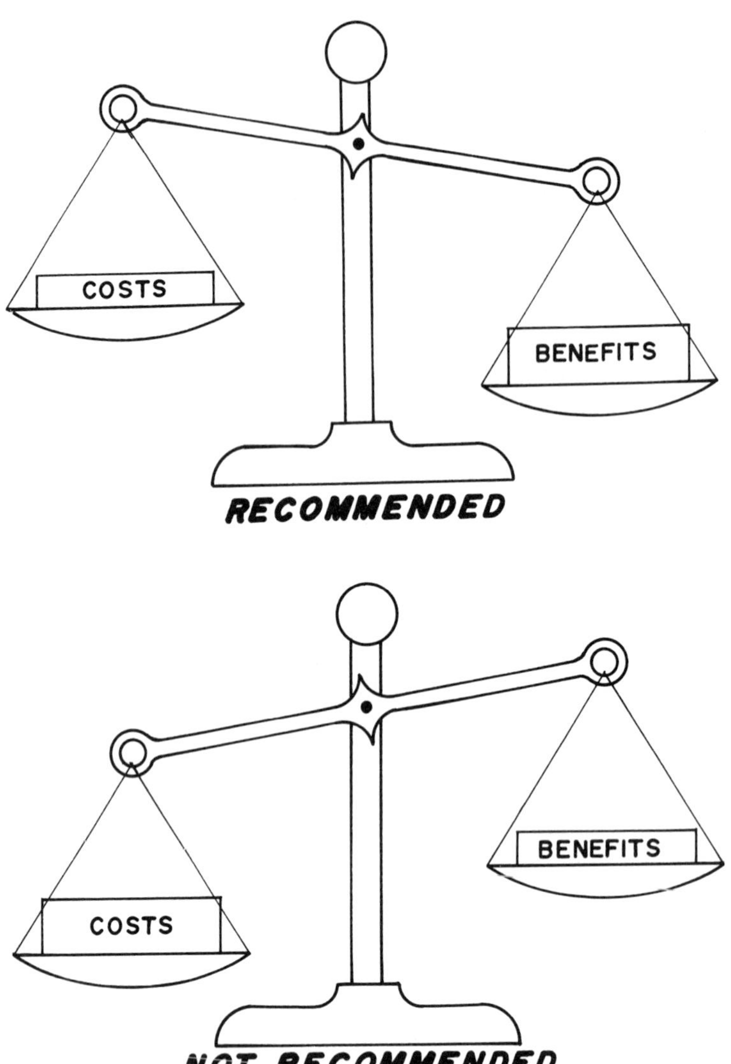

Fig. 9.2 Comparing savings with cost.

Solution:

$$\text{Payback period} = \frac{\text{Installed cost}}{\text{Net annual savings}}$$

$$= \frac{14,500}{18,000 - 3,200}$$

$$= \frac{14,500}{14,800}$$

$$= \underline{\underline{0.98 \text{ years}}}$$

Rounding off $\underline{\underline{1.0}}$ year.

The solution is graphed in Figure 9.3.

9.3.2 Present Worth

The present worth of the discounted cash flow from the proposed investment may be used to evaluate the system. Estimated cash receipts and disbursements are discounted to present value by using compound-interest factors at an appropriate interest rate. The alternative with the largest positive present worth is best using this criterion. This method is often used to evaluate the purchase of an enterprise. It can't be used to compare alternatives with different service lives.

Example:

A company is investigating buying a warehouse. Two locations are under consideration. The required rate of return is 25% per year before taxes

	Location	
	A	B
Land cost	$ 500,000	$2,000,000
Building cost	$2,500,000	$2,000,000
Service life	20 years	20 years
Salvage value after 20 years	$1,000,000	$3,000,000
Operating costs per year		
Materials handling	$ 750,000	$ 500,000
Heat, light, taxes, insurance	$ 100,000	$ 150,000

Which location do you recommend on a present-worth basis?

Symbols:

P = Present worth or investment now

A = Annual amount

L = Salvage value

n = Service life in years

i = Rate of return

Compound rate of return factors used:

1. Present worth of a uniform annual amount = $\dfrac{(1 + i)^n - 1}{i\,(1 + i)^n}$

2. Present worth of a future sum = $\dfrac{1}{(1 + i)^n}$

The numerical values for these factors can be obtained from tables in an engineering economy or mathematics of finance book. They rarely calculated directly for a manual solution. However, most computer programs for economic evaluation calculate the factor directly. Given the operating speed of a modern computer, direct calculation is done almost instantaneously at insignificant cost.

Solution:

	Location	
	A	B
Land	\$ 500,000	\$2,000,000
Building	\$2,500,000	\$2,000,000
Present worth of annual operating cost		
Materials handling		
Location A: \$750,000 per year × PW series 25%, 20 years (3.9539)	\$2,965.500	
Location B: \$500,000 per year × PW series, 25%, 20 years (3.9539)		\$1,977,000

	Location	
	A	B
Other expenses		
Location A: $100,000 (3.9539)	$ 395,000	
Location B: $150,000 (3.9539)		$ 593,100
Subtotal	$6,360,500	$6,570,100
Less present worth of salvage value		
Location A: $1,000,000 × PW of future payment, 25%, 20 years (0.0115)	$ −11,500	
Location B: $3,000,000 (0.0115)		$ −34,500
Total	$6,349,000	$6,535,600

Location A is recommended. The solution is charted in Figure 9.4.

9.3.3 Equivalent Uniform Annual Amount

In this method, annual net savings are compared to the annual equivalent costs including a stipulated rate of return. If the annual savings exceed the annual costs, the proposal meets this economic criterion.

Example:

A materials handling system economic analysis provides the following data:

Installed cost (P) = $44,000

Net annual operating savings (A) = $19,500

Anticipated service life (n) = 6 years

Estimated salvage value at the end of useful life (L) = 0

Minimum required rate of return on investment (i) = 25% per year

EUAA = Equivalent uniform annual amount

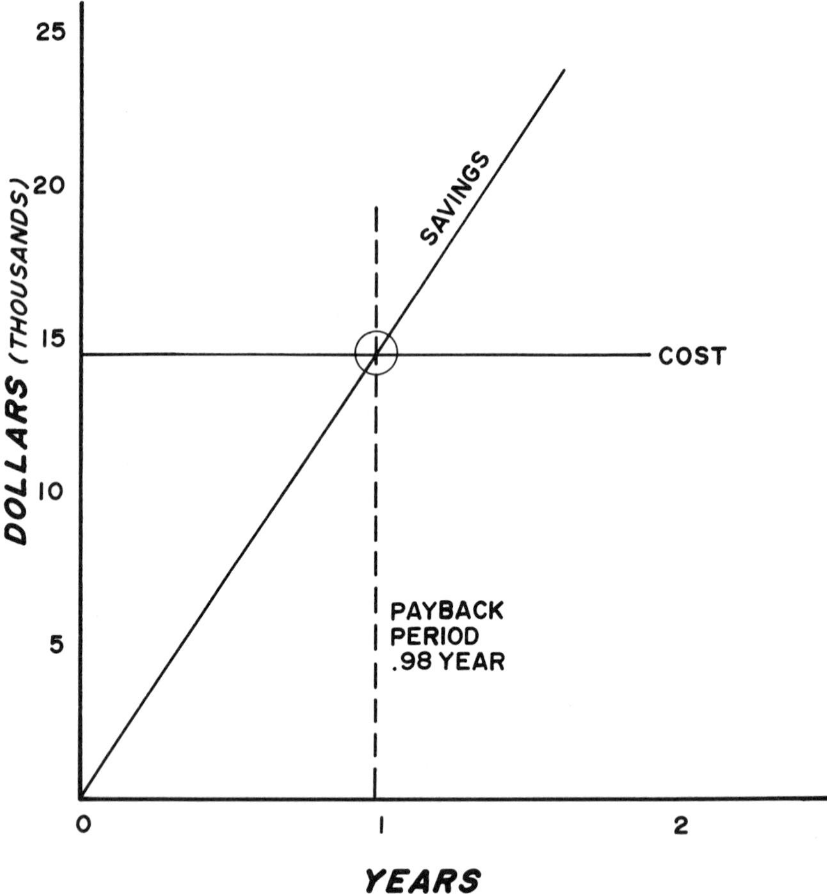

Fig. 9.3 Payback period example.

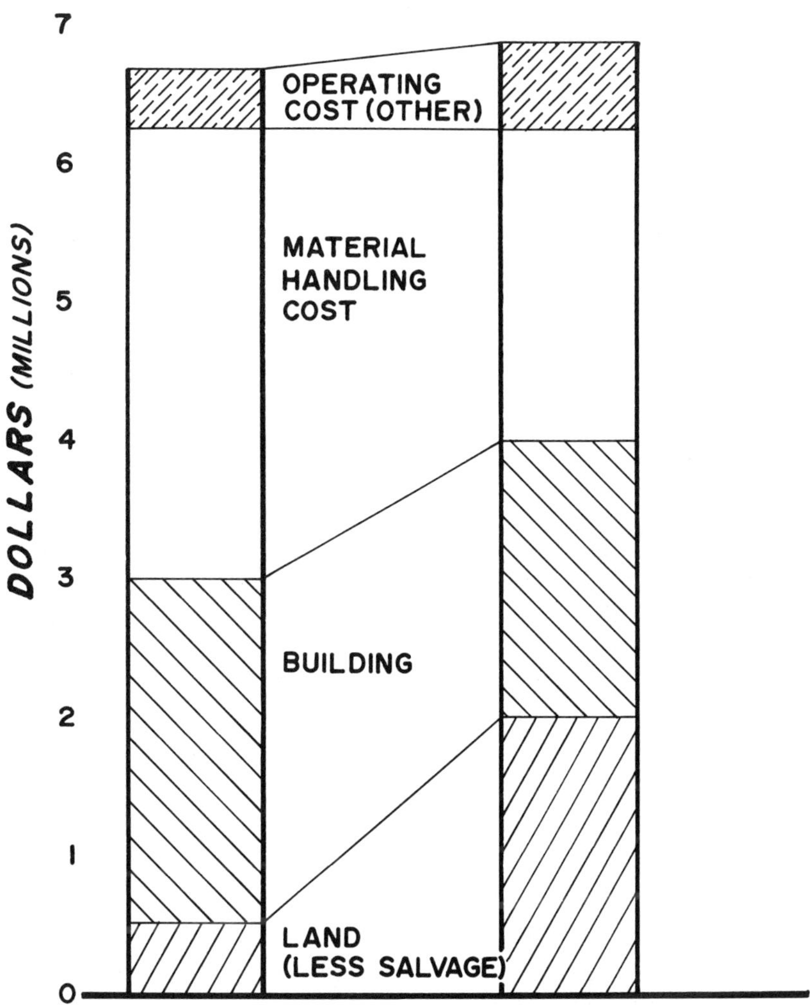

Fig. 9.4 Present-worth example.

What is the equivalent uniform annual profit on the project?

EUAA = P (Capital recovery factor for 6 years at i% rate of return) − (L) (25%) + A

EUAA = −44,000 (0.33882) − 0 + 19,500 = +19,500 − 14,908 = +4592

Since the equivalent uniform annual amount is positive, the project will earn at least the required 25% annually on the investment. It meets the criterion of economic feasibility using this method.

9.3.4 Rate of Return

The rate of return on investment is another basis for comparing alternatives. It can be used for completely different alternatives such as comparing a materials handling system with a building renovation or a new production machine. The return may be either increased income or reduced costs.

Many companies use this criterion in the annual capital budgeting and allocation program. It provides a common measure for evaluating all types of capital expenditure proposals. A typical yardstick is 20% return on investment before taxes. This figure has been used for years by companies such as DuPont and General Motors.

In this method, estimates are made of the annual benefits and expenses, the capital investment, the expected service life, and the salvage value at the end of the useful life. From these, the rate of return is calculated as an annual percent return on the capital investment.

Example:

The data are the same as in the previous example. However, we shall solve for i%, the rate of return on the investment.

0 = A − P (capital recovery factor for 6 years and unknown i%) + (CF)(i%)

0 = 19,500 − 44,000 (CRF)

(CRF) = 0.44318

Interpolating to get i%:

(CRF) @ 35% = 0.41926

(CRF) @ 40% = 0.46126

$$i\% = 0.35 + (0.05)\left(\frac{0.41926 - 0.44318}{0.41926 - 0.46126}\right)$$

$$= 0.35 + (0.05) \left(\frac{-0.02392}{-0.04200}\right)$$

$$= 0.35 + (0.05)(0.570)$$

$$= 0.35 + 0.0285$$

$$= \underline{\underline{37.85\%}}$$

The project will earn 37.85% on the investment.

9.3.5 Unit Cost

The unit cost basis can be used for comparison and for managerial decision. The estimated annual costs are divided by the estimated annual volume to obtain the unit cost. The annual costs are computed by the method shown previously under equivalent uniform annual cost. This method has the advantages of being compatible with accounting unit cost analysis and of fitting into managerial perceptions. Many executives think in terms of the unit cost of each item for each item for each component of material, labor, and factory overhead.

9.4 MULTICRITERIA EVALUATION

Although cost and economic criteria may be the most important and the deciding factors in selecting a materials handling system from among several alternatives, other criteria may enter the decision. Some may be expressible in dollar terms and are included in the economic analysis. Other are nonquantitative, often known as intangible or irreducible factors.

If a decision involves factors expressed in different units or in qualitative terms, the evaluation method must have a consistent rational methodology which includes all pertinent factors in the decision process. One method is to ignore noneconomic factors: this may leave out vital considerations. The oldest and still most common way is managerial judgment. There are several formal methods which organize and present the data in a way that facilitates arriving at the correct decision.

Choosing an airport location for a large city involves several factors, including capital cost, operating expense, safety, convenience to the traveling public, noise pollution, and air clearance restrictions. The first two can be expressed in dollars. Safety is measured in accident and fatality rates, convenience by travel time to and from the airport, noise pollution by decibels, and air clearance by the extent of building restrictions required. Multicriteria evaluation methodology combines these measures to facilitate a decision.

9.4.1 For and Against

One of the simplest methods starts with a list of the advantages and disadvantages of each alternative for each criterion. No attempt is made to weight, rank, scale, or otherwise quantitatively evaluate the factors. The materials handling systems designer and the manager must mentally evaluate the information listed to arrive at a decision.

Example:

The tabulation shown below can be expanded to any number of criteria desired:

	Alternative A		Alternative B	
Criterion	For	Against	For	Against
Flexibility for future modification	Can be rebuilt over a week end	Sacrifices some strength	Better current operating characteristics	Would require 2 full weeks to rebuild

9.4.2 Scaling

One way to combine several criteria into one figure of merit for evaluation is scaling. The estimated worth of each criterion is expressed as a numerical scale value usually on the basis of 1 to 100. The scale values can then be combined since they are independent of the units of measurement.

Major difficulties with this method are establishing the scale and assigning values to the individual alternatives. For example, establishing a scale value for the flexibility of an alternative involves uncertainties of the flexibility evaluation itself and the difficulty of assigning a scale number to the evaluation.

Example:

The scale values for each of four factors are estimated on the basis of 1 to 100. They are tabulated below without weighting for relative importance.

	Alternative			
Criterion	A	B	C	D
Rate of return	90	75	60	68
Payback period	60	66	55	80

Criterion	Alternative A	B	C	D
Flexibility for future modification	65	45	60	72
Lead time to installation	20	85	80	60
Total	235	271	255	280

Alternative D would be recommended.

9.4.3 Ranking

In this method, each alternative is ranked against the others for each of the chosen criteria. The rankings are added. The lowest alternative is ranked 1. Others are ranked in order up to the highest, which is given a rank equal to the number of alternatives. The alternative with the highest rank total is recommended.

Example:

Criterion	Alternative A	B	C
Rate of return	3	2	1
Payback period	2	3	1
Flexibility for future modification	3	1	2
Lead time to installation	1	3	2
Total	9	9	6

Alternatives A and B are equal.

9.4.4 Weighting

Not every criterion in a multicriteria evaluation is equally important. Normally economic factors will predominate. One way to handle this problem is to weight each factor in accordance with its relative importance. Weighting usually goes with ranking or other similar methods.

9.4.5 Factor Analysis

Factor analysis is a combination of ranking or scaling with weighting. A weight is assigned to each factor according to its importance. This weight is then multiplied by the rank assigned to the alternative for that factor. The weighted ranks are added. Note that in this method, the ranks are from 3 for the best to 1 for the worst.

Example:

		A		B		C	
		raw	weighted	raw	weighted	raw	weighted
Criterion	Weight	Rank	Rank	Rank	Rank	Rank	Rank
Rate of return	10	3	30	2	20	1	10
Payback period	8	2	16	1	8	3	24
Flexibility for future modification	5	3	15	1	5	2	10
Lead time to installation	6	1	6	3	18	2	12
Total			67		51		58

Alternative A would be recommended.

9.4.6 Multiple-Criteria Models

Multiple-criteria analysis is a methodology for evaluating alternatives using a model based on utility theory. This theory depends on a factor of managerial judgment of the value of a given level of a measure of merit criterion to the organization. For example, a satisfactory return on investment may have more utility to the decision maker than a much higher return which involves risks regarded as unacceptable.

Multiple-criteria analysis has proven valuable in major decisions involving several factors. However, its use is limited by its complexity, by the amount of time required, and by communications difficulties between the analyst and decision makers.

9.5 NONECONOMIC FACTORS

Some evaluation factors are difficult or infeasible to express in purely economic terms. Some may be quantitative, such as operating capacity. Others may be intangible, for example, environment and safety. Sometimes the cost and difficulty of quantifying a measure may lead to substituting a verbal or nonquantitative surrogate.

9.5.1 Capacity (Volume)

Will the proposed materials handling system be able to transport the product at the rate required by the factory operations? Can it do this at a steady rate which will facilitate dependent operations? Can it handle peak period rushes if necessary?

Data for this factor will come from several sources, including the system manufacturer's specification sheets and representatives, from past experience, and from the engineer's own calculations. The designer should verify the accuracy of the data by using different sources. He should also find out the possible error (standard deviation) inherent in the data.

Volume calculations must include an allowance for downtime due to emergency repairs and other causes. Future increases in volume may pose a problem. Among possible solutions are provisions in the original design for modification to increase the volume or additional working hours.

The variability of the volume flow is important. Some materials handling systems operate at a constant rate. An example is an assembly line. Others have a regular pattern of fluctuation. In many distribution operations, the first hour of the day is light; then volume builds up as the day progresses. The volume in others varies at random when operations are controlled by customer demand, which may follow a random pattern.

9.5.2 Ability to Handle the Product

Can the system handle the product size, weight, and shape? Can it pick up and set down the product efficiently?

The designer should use an appropriate factor of safety. If the system is designed too closely to the minimum engineering requirements, trouble may arise if the design limits are temporarily exceeded or if the product is redesigned to be heavier or harder to handle. The designer must balance lower costs of less rigorous design against the benefits of a bigger engineering design margin.

In one installation observed by the author, tool changes in a flexible manufacturing system were handled by a robot. The robot handled the tool changes accurately without damage. However, it took the robot 5–10 times longer than a machinist or machine operator would.

9.5.3 Maintainability

Maintainability is the measure of the ease and efficiency with which a materials handling system can be maintained. The increased efficiency and lower unit cost of mass producing original equipment have led to lower first costs in relation to maintenance costs. Conversely, higher labor costs, increased scarcity of skilled workers, and the very high cost of lost production, and high carrying cost of spare-parts inventories have brought a substantial increase in maintenance expense. Management today insists that equipment be easy and inexpensive to maintain.

Good data on maintainability are hard to come by. Manufacturers of materials handling equipment may not have the information or may be reluctant to divulge it to a prospective customer. The best source is the company's own experience with similar products from the the same vendor. A second possibility is the opinion of other users of the equipment.

Some maintainability features are discernible from inspection. Electrical parts and oil filters should be easily accessible and fit standard maintenance equipment. Covers should be easy to remove and replace. Controls and gages should conform to good human factors and engineering principles.

9.5.4 Reliability

Reliability is the mathematical measure of the dependability of equipment. It may be measured in percent downtime or its converse, percent uptime. Either way measures the availability of the equipment to do the job it was purchased to do. Since World War II, the engineering science of reliability has developed and spread from military to civilian application. Figure 9.5 plots downtime and reliability costs in a typical situation.

Reliability is another parameter for which good data are difficult to obtain. There are several reasons for this:

1. The manufacturer may not have the data.
2. Good data require a long period of service, which is seldom available for new designs.
3. The manufacturer may not want to release reliability data.
4. Reliability data are usually an average derived from a sample of a number of items. An individual item within the sample may deviate substantially from the mean of the group.
5. Reliability is also a function of the user's maintenance and operations policies. Poor maintenance and continual overloading will decrease reliability.

9.5.5 Damage and Safety

A major consideration in a materials handling system is minimizing the possibility of damage to the product and accidents to employees. These can cause major financial losses. Damaged product with resulting delivery delay loses customer goodwill and repeat business. Accidents with accompanying pain, distress, loss of work time, and disruption are expensive and distressing.

At one time, there was a movement to automate materials handling in furniture manufacture and distribution. However, most automated

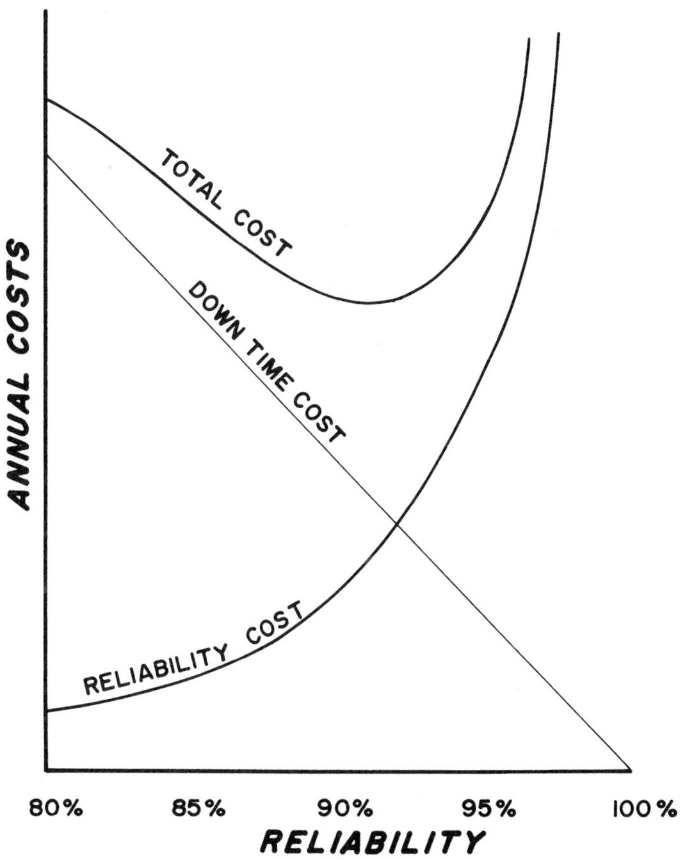

Fig. 9.5 Reliability versus cost.

systems failed and were replaced by manual handling. The major reason for failure was damage to furniture by automated equipment. This damage so reduced the market value of the furniture that the materials handling savings were more than wiped out.

Since the passage of the Occupational Health and Safety Act, workplace safety has been required by law. Numerous regulations have been issued and more will come. Most were adoptions of existing standards and well-known principles of safe design and operation. Many concern the design and operation of materials handling equipment.

9.5.6 Compatibility

Is the system compatible with interfacing materials handling systems and production equipment? Does it fit into the building structure and meet structural requirements and building codes?

A common fault in materials handling system design is suboptimization. This is optimization of the design, operations, and cost of a materials handling system without regard to the overall efficiency of the company. For example, top efficiency in operating the materials handling system may interfere with efficient operation of production machinery. Transfer between systems may not be properly automated and require manual handling. Suitable provision for breakdowns may be omitted, causing an important production operation to shut down if the material handling fails.

The materials handling system design must comply with applicable building codes and zoning regulations. Some cities and states are undesirable and uneconomic locations because of unduly restrictive codes and arbitrary interpretations.

9.5.7 Installation Lead Time

Can the proposed system be designed, acquired, and installed by the date it is needed?

The lead time to completion date may be a controlling factor in materials handling system design. The designer should ascertain through the purchasing department the approximate delivery lead times for key components. Availability may determine system element choices, particularly if the deadline is close enough to create pressure. Short lead times encourage the use of standard, off-the-shelf components and reliance on large, experienced suppliers with adequate resources and a strong organization.

9.6 OTHER CRITERIA

Several criteria for evaluating a materials handling system are neither purely economic nor purely intangible. Most correspond to principles of good design. Some economic aspects are not easily measured in dollars. The following can be easy and quick measures of the system.

9.6.1 Number of Handlings

A basic principle of good materials handling is to reduce the number of times the product is handled. This is based on the assumption that each handling costs money and thus the fewer handlings, the better. Although this may not always be true, the designer and decision maker can use this for a quick judgment of alternatives.

9.6.2 Distance Moved

The total distance moved is another good measure of a materials handling and plant layout system. Again, this is based on the design principle of minimizing the total move distance and the assumption that cost is proportional to distance times volume of product moved.

9.6.3 Smoothness of Flow

This is a more difficult measure. It is based on experience and subjective judgment. Smoothness of flow means avoiding backtracking, minimizing bottlenecks, and moving materials rapidly through the plant. One rule-of-thumb used by some top production executives is "Once the material starts through the plant, it must never touch the floor until it reaches the shipping department."

9.6.4 Inventory

Another way of evaluating materials handling alternatives is the amount of inventory needed to meet production and sales requirements. This is important since each dollar of inventory costs the company about 25–30¢ per year in carrying costs. Lower inventories free funds for other purposes. The amount of money saved through inventory reduction can be included in the economic evaluations described earlier. However, this is not always done in the economic feasibility study. If not, it can be examined as a separate measure of value.

9.6.5 Number of Personnel

The layout that requires the fewest workers is regarded as the best. This is increasingly important as the cost of labor rises sharply. A major component of labor cost is the fixed cost per worker regardless of time worked. This fixed cost includes vacations, holidays, pensions, medical insurance, and life insurance. There is strong motivation to keep down the number of regular employees by automation or by subcontracting.

9.6.6 Extrasensory Perception or "Hunch"

Some executives choose an alternative layout on the basis of "feel," extrasensory perception (ESP), or hunches. Usually, this is a poor way to make an important decision. However, engineering research has shown that some executives have the subconscious ESP ability to arrive at a sound decision without being able to describe the reasons or the thought processes upon which the decision was based.

REFERENCE

1. By permission. From Webster's Ninth New Collegiate Dictionary © 1986 by Merriam-Webster, Inc., publisher of Merriam-Webster ® Dictionaries.

10

Presenting and Selling Your Materials Handling System Recommendations

10.1 INTRODUCTION

A good materials handling system design is worthless unless it is installed and implemented. This will not occur without the approval and support of the manager in charge. The designer must convince the key manager that the design is the best possible under the constraints imposed and that it is to his advantage to approve and go ahead with the system. This requires that the designer be a salesperson and an advocate of the system.

10.2 METHODS OF PRESENTATION

The proposed materials handling system and the designer's recommendations can be presented in one or more different ways. The decision maker usually specifies the method of presentation and the format of the report. Some prefer a comprehensive written report; others prefer a short summary of the important points. Admiral Ernest King, Chief of Staff of the U.S. Navy during World War II, insisted on brevity. Every report longer than one page was returned for rewriting. Some managers prefer oral presentations. Company manuals and policy may prescribe some features of the report, such as the format and distribution.

10.2.1 The Written Report

A written report may be either the major presentation vehicle or a supporting document with material needed by those involved in the project. The form of the report is usually defined by company policy and manuals. If not, the materials handling project director must design the report to best accomplish his objectives.

Time and space do not permit us to include detailed text on report writing. The reader should use a good reference on the subject. Several are available. Some companies have an editorial section from which advice and help on writing a report can be obtained.

10.2.2 Oral Presentation

A manager may prefer an oral presentation either for himself alone or for a group including himself and his staff. This may be the major presentation with a period for questions and discussion.

Careful organization and preparation for an oral presentation are essential to success. The designer and his associates should organize the presentation, prepare the script carefully, and practice. The designer and his staff should review the presentation for possible questions and objections so they are prepared to answer them fully and accurately.

Every engineer should have some training in public speaking and oral presentation as part of either a formal academic course, an in-house or noncredit workshop, or a self-help group such as Toastmasters. Effective public speaking is important to acceptance of one's designs and to personal advancement.

10.2.3 Audiovisual Aids

A wide range of audiovisual aids is available to the designer for his oral presentation. The choice of media depends on the preferences of the designer and the manager, the availability of equipment and services within the organization, and the importance and nature of the presentation.

A detailed discussion of audiovisual media is beyond the scope of this book. The reader should refer to the audiovisual section of his organization or to books available in most public libraries. Many companies and libraries maintain an audiovisual section which can help with advice and references.

10.2.4 Scale Models

Three-dimensional scale models are expensive, yet useful in some presentations. They are good for visualizing three-dimensional systems which could have vertical interferences. They can be worth the cost for important presentations involving outside funding sources or large-scale expenditures.

10.3 YOUR AUDIENCE

The first step in a good presentation is audience analysis. Who will receive your presentation? What actions are you aiming to get? In an audience of several people, who are the key decision makers and who has a veto power? Who is in a supporting role and who will not participate in the major decisions?

Next, what are the audience's points of view? Are they primarily concerned with cost? With efficient operation? With increased production? With a larger share of the market? Staff will have other points of view. The personnel department looks at the effect on skill requirements and manning tables. Plant engineering will look at installation and maintenance.

10.3.1 The Managerial Decision Maker

The most important person in the audience is the decision maker. He is the person whose approval is essential and whose requests for modifications and changes must be honored. If he remains unconvinced, the approval of everyone else is irrelevant. The designer should keep the decision maker's motivations, attitudes, and instructions in mind all through the design and presentation.

10.3.2 The Line Supervisors

The persons responsible for operating your materials handling system are the line supervisors—the foremen and foreladies, the second- and higher-level supervisors, and the production managers. They will look at the proposal from the point of view "How will this affect my job and my operations?" They must be convinced that the proposed system is an improvement and will help their operations. If they remain unconvinced, they may be able to block the system. If they can't block the installation of the new system, they may create trouble by failure to cooperate.

10.3.3 The Staff

Staff are persons who perform functions connected with the installation and operation of the proposed materials handling system, but do not have line responsibility for production or operations. Cooperation of staff will facilitate smooth system installation and operation.

Staff includes, but is not limited to, the following:

1. Personnel. This department keeps track of skill requirements, job classification, pay grades, manning requirements, and similar functions.
2. Accounting and cost accounting. The accountants keep track of the expenditures of funds and the allocation to the company's different projects, products, and functions. They may have an advisory role on aspects of the system.
3. Purchasing. The purchasing agent and his staff are responsible for procuring equipment and services from outside vendors on the best terms needed to obtain the required quality and delivery.

10.3.4 Plant Engineering

This is the organization that maintains the building and equipment, installs and tests new equipment, and works with subcontractors on building and equipment contracts. Its personnel includes persons with the skills necessary to carry out its function. However, large or specialized jobs beyond the scope of in-house personnel are subcontracted.

The materials handling system designer must work closely with the plant engineering personnel. They will provide drawings, data, and specifications on the existing building and equipment. This is essential data since the new system must fit into the present facilities efficiently at minimum expense. Plant engineering personnel can be helpful in suggesting ways to improve the installation and reduce the cost. Finally, the plant engineering department must install the new system and supervise the subcontractors. It is common sense to consider the needs and abilities of plant engineering since they are heavily involved and are in a position to help or inhibit the installation.

10.4 CHARACTERISTICS OF A GOOD PRESENTATION

Presentations differ in format, content, and media. However, every effective presentation has characteristics that separate it from a routine one. A few key points are mentioned here. There are many good books in the field for further reading on the subject.

10.4.1 Speak to Your Audience's Motivations

In the previous section, we discussed your audience composition and characteristics. A basic principle of effective presentation is to speak to the audience's motivations. Speak to cost reduction, to improving production, to better labor relations—to whatever the audience is interested in. No matter how enthusiastic you are, the audience will not accept your ideas and become enthusiastic unless you appeal to their motives.

Don't forget personal as distinct from organizational motives. Cost reduction is a valid company objective. However, if your proposal calls for reaching that objective through reducing or eliminating the positions of those involved, the reaction will be negative no matter how good the proposal.

10.4.2 Answer the Audience's Problems and Questions

A common mistake is to emphasize problems and questions that concern the system designer and project manager. The audience doesn't really care unless these concerns affect them. Be prepared to answer their problems and questions preferably in the original presentation. The designer should have a good idea of what these questions are through discussions and meetings before and during the project. If answers are not included in the original report, the presenter should anticipate them and be prepared to respond during the question and discussion period.

10.4.3 Clarity

Review the presentation to make sure that your ideas and meanings are clearly communicated to the audience. The designer and his staff are too close to the project to realize that what is perfectly clear to them may not be at all clear to others. A separate view of the final report draft from the point of view of the audience is a good procedure. It's hard to take this point of view at the time the design process is underway.

Look for possible misunderstandings. One top utility executive commented, "It's not hard to write a manual that can be understood; it's almost impossible to write one which cannot be misunderstood."

10.4.4 Simplicity

Effective presentations are simple. They eliminate extraneous detail and concentrate on the key issues. Detail not essential to the main presentation can be relegated to the written report, possibly to the appendix.

10.4.5 Visualization

Help your audience visualize your design ideas. Sketches, drawings, and charts communicate understanding more effectively than do verbal presentations. Try to get the audience to visualize your system and its operations and how it will help them in their jobs.

10.5 SELLING YOUR RECOMMENDATION

Some engineers and designers don't like the connotations of the term "selling." However, human activities involve dealing with other people and convincing them to accept your ideas and to follow the course of action you're advocating. This is selling.

10.5.1 Ideas Are Worthless If Not Accepted

The best ideas in the world are useless if the decision maker does not accept, approve, and put them into action. History is full of inventions, discoveries, and ideas that lay dormant for years because they were not publicized and sold by the originator. The goal is not to design a good system; it is to design and put into operation a good system. Selling the idea is essential to getting the system into operation and use.

10.5.2 Your Are a Salesperson, Like It or Not

The only person who can sell your ideas and design is you. The decision maker and his staff members are not going to do this task for you. It is a responsibility that cannot be delegated. If a subordinate makes the presentation, the audience will ask itself, "Why isn't the project manager up in front?" No matter what the answer, the effectiveness will be reduced.

10.5.3 Effective Selling

Effective selling is not an inherited trait; it can be learned. Even though some persons are more effective at selling than others, everyone can and should learn the basic elements of salesmanship in order to be effective.

11

Implementing the Materials Handling System

11.1 INTRODUCTION

Implementation is defined as "to give practical effect to and ensure actual fulfillment by concrete measures" (1, p. 604). In this phase, the approved materials handling project is transformed from a set of drawings and specifications into a physical operating system which moves material. The best plan in the world doesn't become useful and profitable until it is implemented and in action. The effectiveness with which implementation is carried out will determine the degree of success attained by the materials handling system. An efficiently operating system is the fulfillment of the engineer's plan.

The designer seldom has complete charge of implementing the materials handling system. Usually, this is done by a supervisor from plant engineering. However, the designer must work closely with him and with those doing the work. Questions arise that only the designer can answer. Unforeseen events or conditions may necessitate design changes.

One person must be charge to avoid the difficulties of divided authority. At the same time, functional assignments such as purchasing and plant engineering must be respected. The project manager and the system designer both need a combination of managerial ability and diplomatic skill to deal with all those whose cooperation and effort is essential.

11.2 ORGANIZATION FOR IMPLEMENTATION

Good organization is essential for effective action of any kind. Materials handling system design and implementation is no exception; the quality of the design, the smoothness of installation, and the efficiency of the resulting system are all dependent on good organization and competent personnel.

This is not the place for an extensive treatise on organization theory and practice. The designer seldom has control of or influence on the organization structure. The reader should refer to other works on general and engineering management.

A typical organization is shown in Figure 11.1. Other organizational structures are feasible and will work. Much depends on the size and objectives of the company and the availability of qualified personnel.

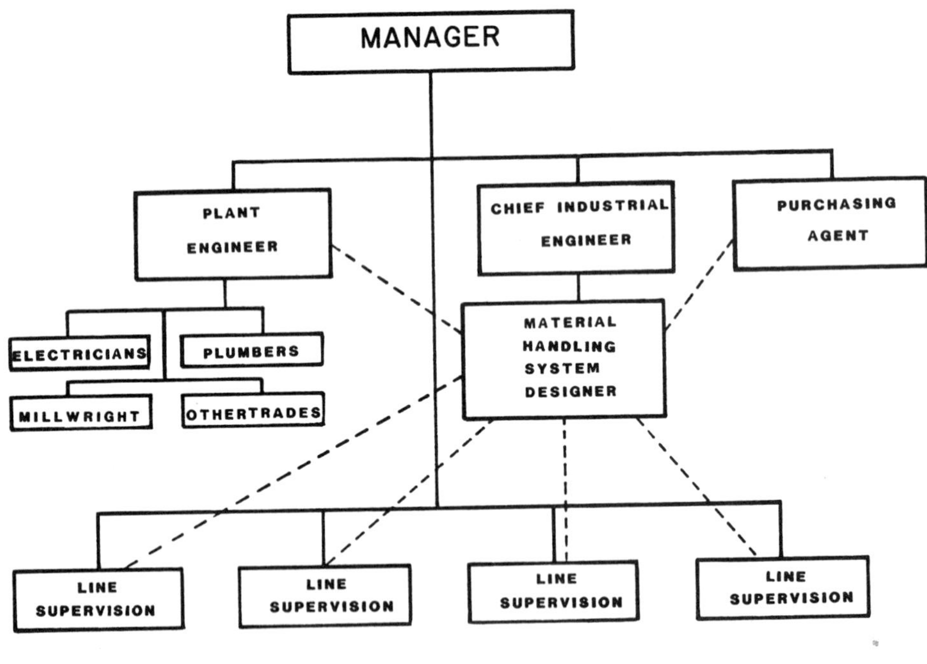

Fig. 11.1 Typical organization for materials handling system implementation.

11.3 ROLES IN IMPLEMENTATION

The materials handling system designer works with many other departments and individuals in the design and installation of a materials handling system. Some of these have a say in the approval or modification of the project or may have the authority to require changes in design and operation. Others may be specialists who can furnish valuable advice on some facets of the system design. The system designer must be able to work with all those involved to secure the best results.

11.3.1 The Materials Handling System Designer

This person is a key individual in the project. He must handle the design function as either an engineer or an engineering supervisor. He must coordinate the needs and requests of the system users and must see that the final output, the materials handling system design, is sound, economical, and efficient. During the implementation phase, the materials handling system designer must be available to answer questions, resolve misunderstandings, and make design changes which are inevitably needed. He is a staff advisor to those in charge of implementation and has no line authority over those doing the work. In some organizations, the designer may also be in charge of installation, but this policy is rare.

11.3.2 The Manager

This is the person who has direct authority over the project and responsibility for its success. Although he doesn't do the design work or make decisions on detail, he is the initiating official and has final approval authority. He must understand the system and its functions enough to make sound judgments and decisions. He is the person who must be the final judge of the project's economic and financial aspects and must be the person who places the project in the perspective of the company's whole operation.

The manager has another important, though not explicitly stated, function. He must show by his interest and actions that the project is important to the organization and must be carried through. This support is necessary to ensure the cooperation and support of other individuals and departments in the organization. If subordinates feel that the boss isn't interested, the project is in trouble.

11.3.3 Line Supervisors

The line supervisor is the person who has to live with the system and make it produce. His ability to make recommendations and request changes will depend on company policy and the managerial style of his bosses. His cooperation is important. Without it, even a good project may fail owing to lack of enthusiasm by line personnel and lack of effort to make the system work.

11.3.4 Plant Engineer

In most factories, the plant engineering department has the function and responsibility for building and equipment installation, removal, and maintenance. Usually the plant engineering department is divided into specialties such as electricians, plumbers, and millwrights. It will also have an engineering design and drafting person or section.

Close coordination between the materials handling system designer and the plant engineer is essential and should start at the beginning of the project. The plant engineer will have vital data on the building, utilities, and other features that influence the material handling function. He can give the designer advice on feasibility of his ideas and may be able to suggest easier and cheaper ways to proceed.

The plant engineer also has the responsibility for coordinating vendor and subcontractor work and for inspecting and approving the work for payment. Many materials handling equipment items are installed and run in by specialized, experienced technicians employed by the vendor. Other construction and installation work will be performed by contractor personnel. It isn't feasible for a company to hire on a full-time basis personnel with all the skills needed for equipment installation and building projects. Wages would be high for low utilization, and availability of scarce skills is limited. In some companies, there are labor contract limitations on the use of outside personnel not in the bargaining unit.

11.3.5 Purchasing Agent

In most companies, only the purchasing department can deal with outside vendors, issue purchase orders, and make binding contracts with suppliers and contractors. The purchasing department will issue requests for quotations, negotiate with suppliers, consult with engineers and others on specifications, and issue the necessary purchase orders. The purchasing department must also be involved in any change orders and modifications to the original contract or purchase order.

11.4 THE IMPLEMENTATION PROCEDURE

After the materials handling system design has been approved and the expenditure of funds authorized, the materials handling engineer is far from through with his project. Someone, sometimes the system designer, has to coordinate carrying out the plan and installing the system. This requires considerable effort, good technical abilities, and personal skills. No matter who is in charge of the implementation, that person must organize and follow a sound plan for successful implementation.

This plan should show what is to be done, who will do it, when it must be completed and what resources are needed. Why it's being done is often omitted. How is left to those doing the work; it is assumed that they have the knowledge and skills needed. If the job is special or unique, some explanation may be necessary.

The plan may be detailed or simple. A large project in a big company may require considerable detail to ensure that all involved are fully informed. On the other hand, a project in a small company may require relatively little documentation. Face-to-face communication can be highly efficient in a small organization.

As shown in Figure 11.2, the steps of the implementation procedure are:

1. Review the approved project for possible changes and modifications and for implementation requirements.
2. Develop an implementation plan. This breaks the project down into logical, manageable parts. It also sets up a schedule for target starting and completion dates. Assignment of jobs and responsibilities to specific departments and persons must be made. The plan should include resource needs in time, personnel, money, and equipment. Decisions must be made on using in-house personnel or outside subcontractors.
3. Arrange for acquisition of system components and services. The purchasing department issues requests for bids, examines the bids, consults with others involved in the project, and issues the purchase orders and contracts.
4. Obtain internal authorizations and work orders.
5. Monitor the progress of the work. The project supervisor must follow the progress of the work and take appropriate action if the work falls behind schedule or if changes must be made or specifications altered.
6. Maintain liaison among the various groups involved. The project director serves as the coordinator and mediator among the various individuals and departments doing the job. Some liaison occurs informally without the supervisor's participation.
7. Inspect and test. As the various subtasks are finished, the supervisor must arrange for testing for acceptance. The actual testing may be the function of another person or department. The director still has to keep track of the inspection and testing and take action if necessary.

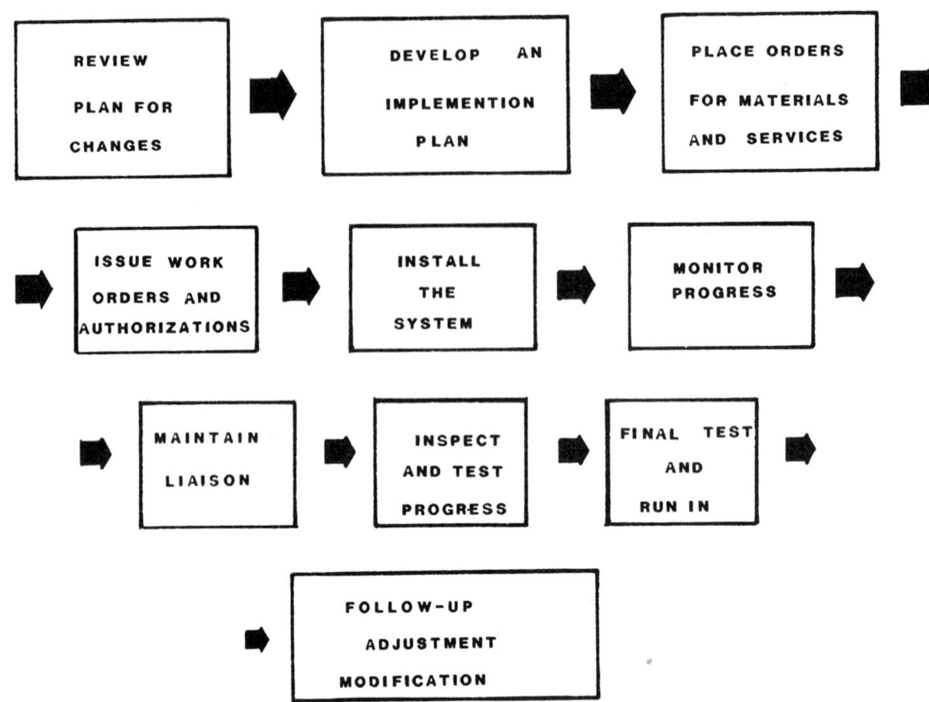

Fig. 11.2 Implementation procedure for a materials handling system.

8. Oversee final test and run in. The project director coordinates the final completion and test of his materials handling system. After this, the system is turned over to those who will operate it.
9. Follow-up, adjustment, and modification. Even the best materials handling system designer cannot foresee all the changes that will be required. After the system is operating, he or she will be called on for corrections and modifications to the system. He or she will have to verify that the system is operating as it should and that the workers understand and are following correct work methods.

11.5 ACQUISITION

Acquisition through purchase or in-house construction is often the controlling time element in a project. This is especially true for items with long delivery lead time or with a long construction cycle. On the other hand, acquisition must wait until the design has been approved and finalized. Starting ahead of finalization leads to expensive change orders since there are almost inevitably changes required from the original design and specifications.

The designer works closely with the purchasing department to ensure that the company gets what it needs on schedule. The designer will be called on frequently to confer with purchasing about technical details on equipment purchases and on subcontracts.

11.6 PERSONNEL TRAINING

Although not necessarily part of project implementation, training of operating and maintenance personnel in new methods and equipment must be provided. This may be a major effort as in the factory introducing robots for the first time. On the other hand, little new training may be required if the system is similar to existing ones and uses the same equipment.

Training is seldom the responsibility of the materials handling project manager. If the company is large, it may have a separate training department. In smaller companies, the line supervisor may do the training. Outside courses, training sessions, or workshops may be used. The equipment vendor may provide in-plant training as part of the sale. In any case, the line supervisor is the key to employee interest and attention.

A major automobile manufacturer is building a new assembly plant in the Midwest. It will be highly automated including a few hundred robots and a number of hard automation machines such as transfer mechanisms. It will be one floor with small columns on approximately 50-ft centers. The company is spending several millions to educate and retrain present workers in the new technologies of maintaining robots and other new materials handling equipment. In addition, the maintenance work force is being reorganized to fit the new technologies.

REFERENCE

1. By permission. From Webster's Ninth New Collegiate Dictionary © 1986 by Merriam-Webster, Inc., publisher of Merriam-Webster ® Dictionaries.

12

Computer Developments

12.1 INTRODUCTION

In recent years, computer capabilities have increased by orders of magnitude and are expected to improve even further. These improvements include more functions and better efficiency together with greatly increased storage capacity. At the same time the cost per unit of function and storage has dropped and computers have decreased in size.

One result of these computer developments has been a revolution in the technology of materials handling. New materials handling system possibilities are available. Automation of the materials handling system is now both technically feasible and economically worthwhile. Real-time reporting and control are possible.

Many computer applications handle low-level, simple, dedicated tasks. No decision making is required. The computer, usually a microprocessor, can be simple, rugged, and reliable.

At a higher level, the computer system becomes more complex as two major functions are added to low-level capabilities. First, the computer participates in decision making. An example is an automatic inspection station. The computer can be programmed to take different actions depending on the preprogrammed limits on the parameter being inspected. The other function is control. The computer system can be programmed to control the operations according to the feedback it receives.

Among the changes accompanying the computer revolution are:

1. Integrated circuits
2. Easier programming and reprogramming
3. Microprocessors small enough to be installed aboard materials handling vehicles.

At the same time, mechanical controls and devices have become obsolete or confined to limited applications.

Another development has been the dedicated computer or one that is dedicated or confined to a single task. Dedication simplifies programming and operation and reduces delays from congestion and higher-priority tasks. A dedicated computer can be located near its task. This improves the physical and organizational communications between task and computer. The dedicated processor can be controlled by a higher-level computer. A later development is utilizing unused computer capacity through a distributed data-processing network.

12.2 DISTRIBUTED DATA PROCESSING

12.2.1 Definition and Description

Distributed data processing is defined as locating computing capability and data at decentralized locations with local autonomy yet keeping centralized corporate control. This ensures that data processing is done at the user location, yet is controlled by and connected to the corporate data-processing center.

Distributed data processing is a system of networking an organization's computers so that functions can be shifted within the network to improve the operations and efficiency of the system. An organization may have several different computers, some dedicated to specific tasks. By redistributing function and tasks, the system raises its utilization of capacity and reduces computing costs.

12.2.2 Capabilities

The advantages of distributed data processing are:

1. Computer capability is assigned to decentralized locations which can benefit by independent processing.
2. Local autonomy can provide for user needs.
3. Databases may be shared within the system. Costly duplication can be avoided.
4. Common applications can be standardized and implemented throughout the entire system.
5. Centralized control provides standard guidelines and

procedures and can hold down excessive expenditures and duplicative effort at decentralized locations.

6. The decentralized control reduces the expense and difficulties of communicating with the centralized host computer.

There have been some problems with industry standards and with compatibility with other computers in a distributed network. Progress is being made toward the goal of compatibility of all computers within the system.

12.2.3 Applications

For example, an automatic storage and retrieval system may have its own system controller computer physically located at or near the AS/RS. The controller issues instructions to the system, plans the routing and scheduling for maximum efficiency, and records in real time the inventory transactions and system status including the location of each AS/RS vehicle. The corporate host computer receives data it needs yet is not involved in operating decisions that are not pertinent to its functions.

Distributed data processing can be organized along divisional, geographical, or functional lines. If the company grants a large measure of autonomy to each operating division, it should also grant computing capabilities, authority, and responsibility consistent with overall corporate objectives and the division's autonomous status. Divisions within the company may use different processes to make different products for dissimilar markets. Consequently, divisional data-processing needs may vary considerably. (See Figure 12.1.)

If a company is organized into regions, it makes sense for each region to have its own distributed data-processing capabilities. This enables the region to design its own system to meet its needs consistent with central guidelines. A major function of centralized guidelines is to avoid duplicating effort in designing and programming systems common to all regions. This is illustrated in Figure 12.2.

Distributed data processing can be effective when organized by function. Each function has different needs. For example, a system for materials handling in an AS/RS bears little resemblance to one used for accounts receivable. (See Figure 12.3.)

12.2.4 Developments

Developments in distributed data processing fall into three catagories: hardware, software, and organization. Computer hardware is increasing greatly in capability and storage capacity at the same time costs and size are going down. Hardware for distributed data-processing centers is becoming cheaper and is often a minor cost element in

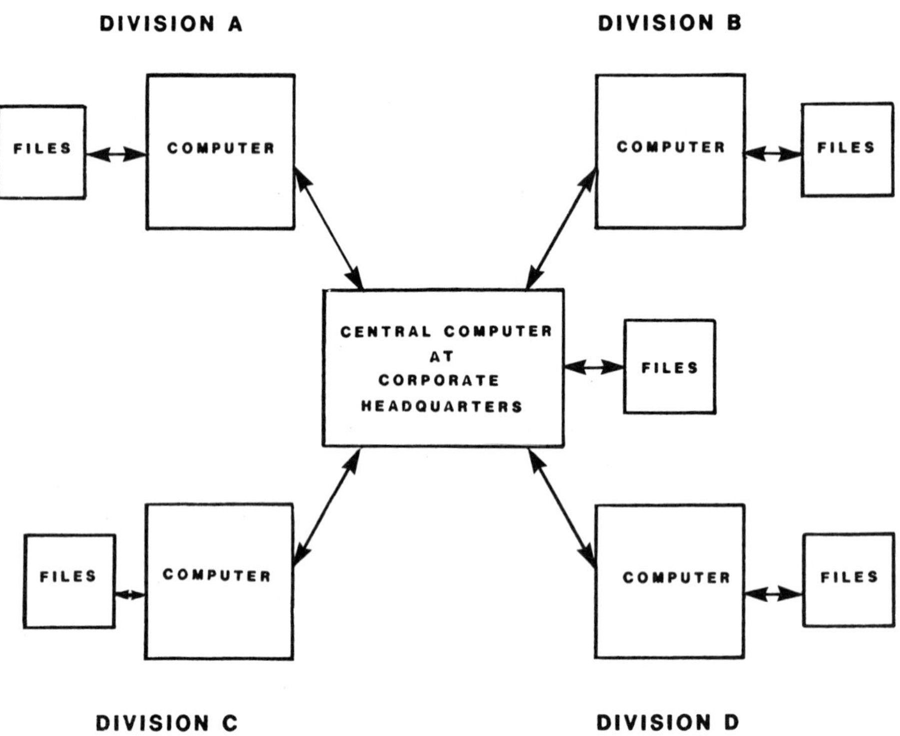

Fig. 12.1 Distributed data-processing organization by division.

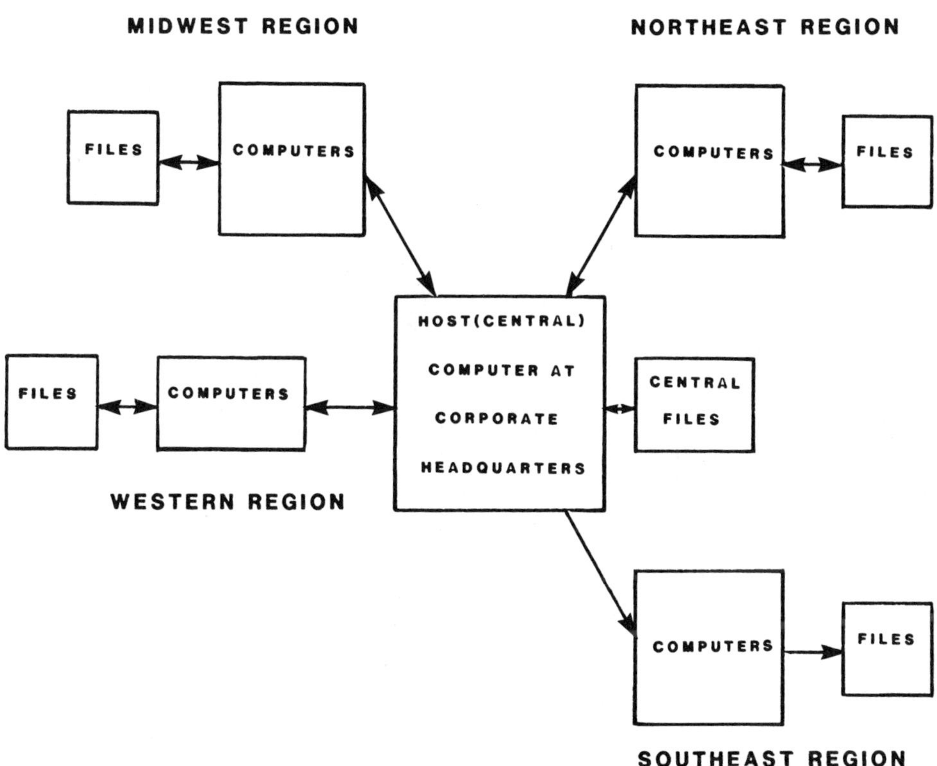

Fig. 12.2 Distributed data-processing organization by region.

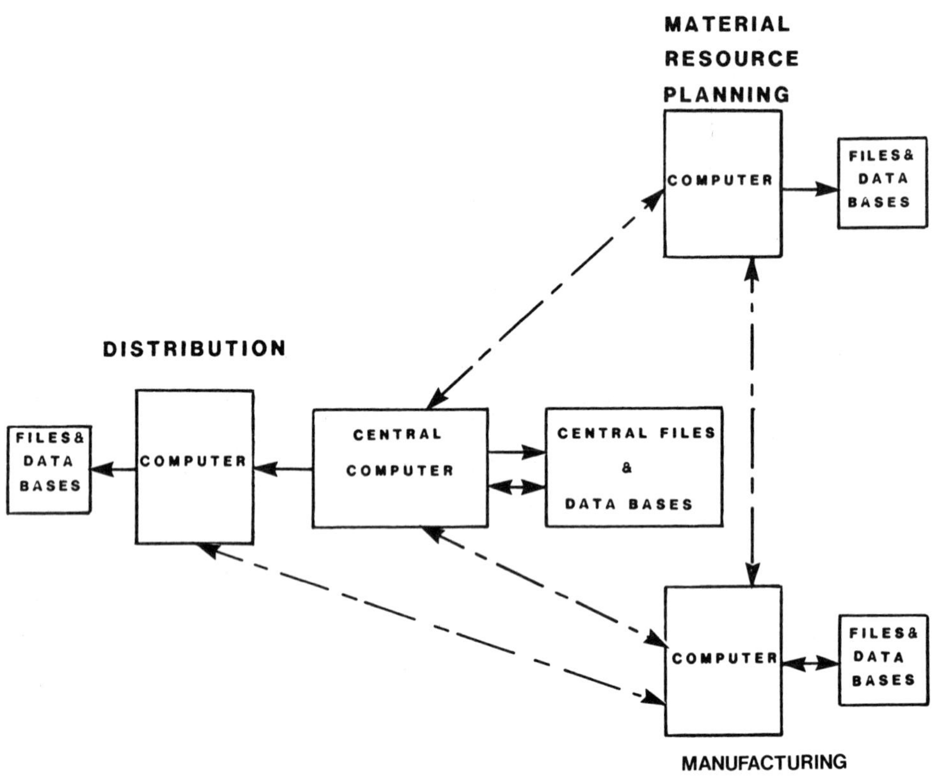

Fig. 12.3 Distributed data-processing organization by function.

comparison to software and organization costs. Software is more powerful and cheaper as distributed data processing becomes more widely used.

A third development is organizational. The central data-processing unit with a monopoly on all the organization's computer functions is passing from the scene. It was too expensive, too slow, too cumbersome and was unresponsive to user needs. Overall, the trend is away from the highly centralized organization to a decentralized structure with greater autonomy and responsibility for the operating units. With organizational decentralization goes computer decentralization. At the same time, top management needs information and control to perform its functions. Distributed data processing meets the needs of both top management and decentralized unit managers.

12.3 HOST COMPUTER

12.3.1 Definition and Description

The host computer (or mainframe) is the large central computer which is the heart of the organization's computer system network. It is characterized by very large storage capacity, high speed, and ability to perform a large number of functions. It handles the major data processing and the overall production planning for the organization.

12.3.2 Functions

A major function of the host computer is keeping the company's records of financial transactions for accounting and other purposes. These form the basis for management reports and for periodic accounting statements. The computer may keep the perpetual inventory by the usual catagories: raw materials, work in process, finished goods, and supplies. It also may keep the master production schedule and issue schedules.

However, the concept of a bigger and better host computer to perform all computer work within the organization never was and never will be realized. The large mainframe is uneconomical for many computer jobs. Communication with the host computer is complicated and subject to delays and errors. The usual personnel organization and structure makes coordination between the host computer staff and the data users and providers difficult. The physical separation of on-the-spot user and host computer poses other difficulties.

The solution to these problems was the distributed data-processing network which decentralized those functions which were better performed at the user location. For example, work-in-process movements on the floor and location of an inventory item in random access storage are no concerns of top management or any other user of the host computer. Consequently, a system controller is installed to manage random access storage and to record location data in an easily accessible format. The minicomputer or programmable controller went even further in decentralization. An example is an on-board controller of an automatic guided vehicle (AGV). This unit handles such functions as correcting deviations from the guide path. This duty is not pertinent to either the host computer or the system controller.

12.4 SYSTEM CONTROLLER

12.4.1 Definition and Description

A system controller is a computer (usually a minicomputer such as a VAX® or PDP®) that manages a subsystem within the organization. An example is the computer that controls the operation of an AGV system. It receives instructions telling what is to be accomplished and collects data on the system status. From these, the system controller issues instructions to the individual AGV. It also receives, collates, and stores reports on AGV movements and completed tasks. It is programmed to send certain data on to the host computer if these items are inputs to the host computer system. Data not needed by the host computer are stored by the system controller until obsolete and then dumped.

12.4.2 Capabilities and Functions

Today's system controller is much more powerful than the largest mainframe of only a few years ago. It is secondary only in the sense that it is subsidiary to the host computer and is dedicated to a specific system function. It can be programmed and reprogrammed to suit system function and has its own storage capacity. It can perform many computer functions only a few of which may be required for system control and operation. It has fault protection usually in the form of redundant circuitry and storage.

12.4.3 Applications

One common application of a system controller is managing an automatic storage and retrieval system (AS/RS). The controller receives instruction on what is to be moved where. This may be to retrieve a pallet load of a particular stock-keeping unit for movement to the shipping dock. The controller locates the required item, recalls the location of the desired item from its memory, selects the vehicle best located to get the load, and issues instruction to the vehicle. The vehicle then proceeds to the item location, gets the load, and takes it to the designated destination. When this task is completed, a report is sent to the system controller that the assigned task has been completed, that the vehicle is in a specific location ready for the next task, and that the storage location is vacant. The system controller records these data, makes the necessary adjustments to its information base, and uses the updated information to prepare the next instructions.

12.4.4 Developments

System controllers are still in the developmental stage. Capabilities and capacities are steadily being improved. Reliability is better and costs and size are decreasingly substantially.

However, greatest potential for improvement lies in software for system operation. Among current developments are:

1. Standard programs for widely used functions
2. More efficient programs
3. Programs that can be modified and updated more easily, quickly, and cheaply
4. Better interactive programs for interfacing other computer systems within the organization

12.5 MICROPROCESSORS

12.5.1 Definition and Description

A microprocessor is a small computer with the following characteristics:

1. Dedicated to a single task such as on-board control of an automatic guided vehicle.
2. Can be positioned at the point of control. This reduces communication time and complexity and also eliminates much of the need for continuously communicating with the system controller at a distance.
3. Can be controlled by a higher-level processor.
4. Some new microprocessors are reprogrammable. Many early ones had fixed programs (so-called "hard wires") and had to be replaced if the function changed. Newer ones can be reprogrammed if necessary, often by substitution of a new integrated circuit chip.

Some examples of microprocessors are digital readout thermometers for hospitals, digital watches, and washing-machine controllers.

12.5.2 Examples

A common example of a microprocessor is the digital readout thermometer used in many hospitals. A transducer converts temperature into a digital signal which the microprocessor converts to a numerical reading. In addition to being much faster than the obsolete mercury scale thermometer, sanitation is improved by using a disposable paper cover for each use.

Another microprocessor operates the home washing machine. It converts the user's dial and button instructions into appropriate commands to the motors and other parts of the machine.

An industrial application is a microprocessor that converts machine tool dimensional readings into digital form. The operator can read the digital display quickly and act accordingly. Accuracy is increased and errors reduced.

12.5.3 Industry Standards and Compatibility

Compatibility of computer hardware and software is a major consideration in procuring these items for a company's materials handling system. For various reasons, the product from one manufacturer will not interface with products of others. Usual industrial practice is to make hardware and software compatible with the products of IBM, the dominant manufacturer. However, this is not always true. In some cases, the system elements can be made compatible through interpreters or special programs. This is expensive and not always satisfactory.

Industry standards for compatibility, specifications, and characteristics are being developed. However, this takes time, especially since computer technology is changing at a rapid rate. Also, a manufacturer may feel that his market position will be hurt rather than helped by industry standards. The consensus required for commonality is difficult at best.

12.6 PROGRAMMING AND SOFTWARE

12.6.1 Introduction

Computer programs generally known as software are the brain of any system based on computers. These furnish the directions that enable the computer hardware to function to produce desired results. Without good software, the best computer is worthless.

Originally, the mainframe manufacturer provided the software as part of a package deal. However, this was ruled out as violating antitrust statutes. Now, software can be obtained from one of three sources: the mainframe manufacturer, a specialized software vendor, or an in-house programming department. Many users procure from all three sources. Purchased software may have lower first cost. However, it may not be suited to the user's application and consequently require extensive modification. In-house programming is time-consuming and expensive in comparison to purchasing a package program. The user must examine each situation to determine which alternative is best for him.

12.6.2 Standard Programs

Standard operating systems are now more generally available. This reduces the need for custom programming, a time-consuming and expensive task. Better software for materials handling systems is coming on the market. This is true for other manufacturing and distribution applications.

The availability of standard software has not completely eliminated the need for programmers and custom programming. A standard program must be modified to suit the user's computer hardware, systems, and operating needs. However, better software decreases the amount and complexity of special programming by the system user.

12.6.3 Reprogramming

Computer programs, like other elements of American industry, are dynamic not static. Changes and updating are required continuously. Conditions change, new technology and methods are introduced, new products added, old ones dropped, and new materials handling systems designed. All these changes call for corresponding modifications in programming and software.

Ease of reprogramming is an important characteristic of a good software system. Some programs are easier to modify than others. One major requirement for reprogramming is good documentation of the original program and of all changes. Unfortunately, program documentation is not noted for clarity of composition and exposition.

Some reprogramming difficulties may arise with proprietary or copyrighted software. The purchaser of restricted software should clearly define in the purchase order the procedures, costs, and rights involved in reprogramming.

12.6.4 Programs That Write Programs

A distant research goal is a computer program that will write new computer programs on command. Some progress has been accomplished, but the goal remains far into the future. Many doubt that it will ever be attained.

12.6.5 Programming Languages

Better programming languages are coming on stream. These are easier to use or "user friendly." The use of older languages such as FORTRAN and COBOL is declining, even though they will not completely disappear for a long time, if every. They are too firmly established and are used by too many programmers.

12.7 FAULT TOLERANCE

12.7.1 Introduction

Fault tolerance is the ability of the computer system to keep operating in spite of errors and technical difficulties. These include disturbances in the electrical system supplying power to the computer, loss of program or data, errors in the system, or mistakes in the input data. Some faults can be disastrous. A line disturbance in the power source can erase large segments of a program or data file. Other faults are less severe.

One major advantage of the human operator over the computer is the ability to detect some errors by inspection. For example, if an input instruction calls for a dimension that is off by two orders of magnitude, the human operator will recognize that the dimension is impossible and act accordingly. If the same data are fed into a computer, they will be processed just as though they were correct. Many protective subroutines and programming steps must be inserted to guard against obvious errors that a human would identify and act upon immediately.

12.7.2 Developments

Computer fault tolerance is improving. The modern system can sustain error or failure without functional interruption. Without fault tolerance, small line disturbances or other events could shut down the system or erase programs and data.

Since one of the major criteria for an efficient computer system is reliability, much attention and effort has been directed toward improving that reliability through improving the system's ability to continue operating in spite of faults, errors, and disturbances. Various methods have been developed. One is redundancy. Programs and data files are duplicated so the loss of one does not shut the system down. If something happens to one program, the system switches over to the redundant program or file and keeps right on operating. Methods have been developed to protect the computer system and files against destruction. Other techniques have been originated to identify program or data errors and allow for or correct them.

12.8 LIMITATIONS OF COMPUTERS

Computers do have some limitations. Among them are:

1. A computer can do only what its program calls for. A distant dream is the computer that can develop its own programs. This dream will be a long time becoming a reality. Maybe it never will.
2. Software development and modification is still slow. Another long-range goal is a methodology for writing and changing software quickly, cheaply, and efficiently.
3. Computers require precise system program design, implementation, and testing. It is not uncommon for a computer installation project to require more time for analysis and improvement of the organization's operations than it does for the computer installation and programming.
4. The speed of light. The upper limit on the speed of transmission of electrical and electronic impulses is the speed of light, 180,500 miles/sec. Although this may not sound like an important limitation, it does limit the number of impulses that can be transmitted within a computer system. One of the advantages of miniaturization is minimizing of the distance that an impulse must travel. This cuts the time required for the signal to reach its destination.
5. Machine vision. This is the ability of the system to gather visual data on system product or elements, interpret the data, identify the item, its location and orientation, and take appropriate action. Much research is underway in machine vision. However, at the present time, system capabilities in this area are primitive and limited.

12.9 MATERIALS HANDLING APPLICATIONS

12.9.1 Introduction

No development in the last few years has produced as many far-reaching changes in materials handling systems as has the improvement in computer technology and computer economics. Improved technology has made automatic materials handling systems technically feasible. At the same time, reductions in cost and size have made these technical developments financially attractive to material handling users.

12.9.2 Operational

One area in which computers have changed materials handling systems is the storage and retrieval data on inventory quantities and location.

They provide real-time status information on where each inventory lot is stored. Also, the location, status, and work ahead of each vehicle are immediately known. This eliminates the thorny problem of inventory which is known to be on hand but which is worthless because it cannot be located.

Computers have automated most record keeping in American industry. This had reduced the number of clerical workers and the number of errors and has provided quick or real-time data service. Considering the amount of data and records needed to operate a modern organization, this is a major productivity advance.

Vehicle guidance and control can now be done with a computer-based system which directs the vehicle from its location to the point of need by the most efficient route. The system can also keep track of vehicle movements and status and provides real-time data on inventory quantity and location.

As a result of decreased size (miniaturization), on-board microprocessors can be placed on materials handling vehicles such as automatic guided vehicles (AGV) to control their movements. The microprocessor can take corrective action if the vehicle strays from its designated path. It can also stop the vehicle on contact with an object or person so no injury or damage occurs.

Flexible manufacturing systems (FMS) automate small-batch manufacturing through a computer-controlled network. The FMS implements quick change of setup so that the machine tool can shift quickly from one product to the next. It also directs materials movement and monitoring. The system is based on a series of computer-controlled centers, which may be machining, inspection or assembly. The centers are coordinated and controlled by the computer system.

Among the characteristics of an FMS are:

1. High-speed operation
2. Robotics wherever feasible
3. AGV for transporting raw material, work-in-process, and finished product
4. Computer-controlled materials handling and storage
5. Easy programming and reprogramming

Routing and dispatching have been computerized with a decrease in cost and an increase in accuracy and efficiency. Computers can provide schedules that will most economically meet production schedules and requirements. Inventories have been reduced at the same time machine utilization has increased and on-time delivery improved.

Automated systems have replaced the human operator in tedious, undesirable, or hazardous jobs. An example is the use of robots in a nuclear environment. This eliminates the exposure of the worker to radiation. Another is the automation of painting. Electrostatic painting equipment run by computers allows the painting to be done in an enclosed space with no human operator. Thus, no worker is exposed to unhealthy paint fumes.

Computer integrated manufacturing links planning and scheduling with the production machinery. It controls the physical movements of the work-in-process and tracks the material throughout the system. A further function is operating and controlling the production machine tools.

12.9.3 Intangible or Indirect Benefits

Several indirect benefits come from the computer-driven improvements in materials handling systems. They include:

1. Real-time control. The system can provide real-time data for decision making. This can eliminate costly delays in making adjustments and allows management and staff to act when the need occurs, not when data come in later.
2. Discipline. The computer-based system requires that the system and those operating it have the discipline to work accurately and perform in a timely manner. The system will not tolerate sloppy work or omissions.
3. Organization. Many of the benefits of a computerized system come from the requirement that the process and system be well organized before it can be computerized. Divided responsibilities, unclear procedures, and other defects must be corrected before the system can be formalized.
4. Process visibility. Design and operation of the system make its elements more visible to management, staff, and operators.

12.10 TRADEOFFS

12.10.1 Introduction

To trade off is "to give one thing in exchange for another" (1). In materials handling system design, it means reducing one element or advantage to gain in another element or advantage. A common example is to trade off lower first-investment cost for higher operating expense. Every system design involves tradeoffs among desirable and undesirable characteristics.

12.10.2 Data versus Control

One common tradeoff is between data and control. A major function of every materials handling system is gathering data on operations, inventory, movements, and status. This takes computer time and memory. On the other hand, the system is expected to direct and control the system. This also takes computer function and capacity. The designer may have to decide which is more important, data or control, and act accordingly. A third factor may enter the tradeoffs. It may be possible to avoid a tradeoff between data and control by increasing computer capabilities through greater capital investment.

12.10.3 Time versus Memory

Another tradeoff is between operating speed and memory. Even though we are dealing with infinitesimal fractions of seconds, the total time required for a computer routine may be important. The time required to process, store, and retrieve data in memory may add up. This total may be at the expense of the speed at which the computer functions in making control decisions and issuing commands.

12.10.4 Speed versus Complexity

Another balance the designer must determine is between speed and complexity. Each program step takes a small, but finite interval of time. If the number of these intervals goes up substantially through adding steps that increase the complexity of operation, the reaction time of the system is slowed. This may or may not be important, depending on system requirements. However, if speed is obtained through simplicity and decreased complexity, the flexibility and versatility of the system are reduced. Again, the designer must determine the tradeoffs in light of system requirements and resource availability.

12.10.5 Closed Loop versus Open Loop

A closed-loop system is one that operates without the intervention of the human worker. An open-loop system depends on a worker for part of its operation. Usually, the function of the human is decision making based on feedback from the system. The open-loop system may be less expensive. Also, there are many situations in which automated decision making (artificial intelligence) is not feasible. This is particularly true if the decision points and rules are fuzzy or the decision rests on a multitude of factors not readily programmable. An example of the latter is a medical diagnostic routine. The physician's diagnostic process is incredibly complex and often depends on his evaluation of factors that have no clear cutoff points or well-defined correlations with other factors.

12.11 COMMERCIAL SYSTEM INTEGRATORS

A commercial sytem integrator is an organization which designs and installs systems involving computers, but which does not manufacture computers or peripherals or write any software. It designs the system to suit the customer's requirements, recommends the procurement of specific system components, supervises the installation of the system, and aids in system implementation, break in, and personnel training.

A noticeable trend at the annual materials handling show and forum (LOGISTEX'86 ®) was the increased number and use of system integrators for materials handling system design and installation. Some are independent consultants while others are a division of an equipment manufacturer or large contracting firm.

12.12 PROGRAMMABLE CONTROLLERS

12.12.1 Definition and Description

The National Electrical Manufacturers Association (NEMA) defines a programmable controller as:

> A digitally operating electronic apparatus which uses a programmable memory for the internal storage of instructions for implementing specific functions such as logic sequencing, timing, counting, and arithmetic to control, through digital or analog input/output modules, various types of machines or processes. A digital computer which is used to perform the functions of a programmable controller is considered to be within this scope. Excluded are drum and similar sequencing controllers (2).

Programmable controllers (PCs) were developed to replace relay controls in manufucturing operations such as materials handling and process control. They have several advantages over relays and have largely replaced them in many operations. In design and operation, they can be regarded as a form of digital computer.

12.12.2 Examples

The first major application of PCs was in the automative industry. There was a need for controls that would reduce the time and cost of the annual model change yet provide the control capability needed to operate a complex, intricate manufacturing system with a minimum of human intervention. PCs provide this capability.

One application is the use of PCs to control the engine assembly line from start to the shipping dock. Another is the PC's control of the robots that weld the auto bodies as they move along the line (3).

Materials handling is a function often controlled by PCs. In one glass factory, product moves from the kiln by conveyor through quality assurance, packaging, and palletizing for shipment under the direction of PCs. The system uses several auxiliary conveyors and diverters (3).

Another reported application is in a large bakery. A major stimulus for switching the PCs was the increasing difficulty of getting replacements and parts for the relay control system (1978). PCs were installed to control the palletizing operation which handled a wide variety of baked products. Substantial savings in maintenance and reliability were expected (4).

12.12.3 Characteristics

PCs are designed to operate in a shop environment that is hostile to computers and computer elements. The usual computer, minicomputer, or even some microprocessors may not operate reliably in the shop owing to contamination, dust, dirt, fumes, heat, and rough handling. The PC is built to stay operational in spite of these hostile influences. It is rugged, tough, and reliable.

Another feature designed for shop floor use is the ease of programming. Many use the relay-logic format which minimizes retraining required for electricians and other shop personnel to perform the programming and reprogramming. These are usually written in shop floor language. The design and operation of the PC are aimed at easy and efficient installation, modification, and use by shop personnel.

Many PCs incorporate self-diagnostic routines. The PC keeps track of its own operations and signals the need for routine or emergency maintenance. PCs also incorporate commonality of control so that integrating several PCs into a single system is feasible.

Input/Output

The input device receives signals from the equipment sensor and transmits them in suitable form to the PC's processing unit. Some are digital, such as photoelectric counters or limit switches. Others can be analog devices; examples are transducers of various types, including pressure, force, temperature, and voltage.

The output device issues the processor's commands which activate or deactivate the operating components or system. The command may be to divert a unit to an auxiliary line, to move the unit to the next station, or even to stop the line for an emergency repair.

A common example is a presence-absence check. If the input sensor indicates that there is no part in a line station, the processor will direct that no further parts be added or operations performed.

Memory

Memory is the part of the PC that stores the instructions and data needed for operations. Memory systems are similar to those used in computers with modifications to suit the PC's task and functions. Each separate instruction or data bit is stored in a location known as an address. Efficient storage of items into memory and retrieval when needed are essential to smooth PC operation.

There are two major catagories of memory. Random access memory (RAM) permits an individual bit to be read into memory and later retrieved an unlimited number of times. Read-only memory (ROM) does not permit changing the bit once it is stored in memory. For example, the daily data and instructions can be stored in RAM while the operating language goes into ROM. A major disadvantage of the RAM is that a power failure, even a short one, can wipe out all information stored in the memory. With ROM, the information bits are stored permanently and cannot be wiped out by power failure.

Various memory devices are available. Some, like transistors and punched cards, have become obsolete. Current types include magnetic core, ferromagnetic and flexible disks, semiconductors, charged-couple devices, and magnetic-bubble memories.

Processor

The processor is the part of the programmable controller that receives the data bits from the input device, combines the input information with instructions and data stored in memory, performs an analysis, and issues commands based on the analysis. In structure and operation it is similar to larger computers. However, it is modified to suit the PC's task. Any computer operations not needed for the assigned task are eliminated. For example, if mathematical calculations are unnecessary, this capability will be left out.

REFERENCES

1. By permission. Webster's Ninth New Collegiate Dictionary © 1986. Merriam-Webster, Inc. Publisher of Merriam-Webster® Dictionaries.
2. National Electrical Manufacturers Association (NEMA) from NEMA Standards Publications ICS3-1983 Part ICS3-304, Industrial Systems, copyright 1983 by NEMA, by permission.
3. Programmable Controllers Find New Applications in Material Handling, *Material Handling Engineering*, Vol. 36, No. 9, (pp.) 77–83 (1981).
4. Now Programmable Controllers Are Economical for Smaller Material Handling Systems, *Material Handling Engineering*, Vol. 31, No. 9, (pp.) 72–75 (1976).

13

Automatic Identification

13.1 INTRODUCTION

13.1.1 Definition

Automatic identification is the determination of the key characteristics of an item or product without human intervention. Among the characteristics so identified are product designation, quantity, source, destination, size, weight, accounting data, and routing. These data can be used for scheduling, routing, inventory, storage location data, production and inventory control, shipping documents, and billing.

13.1.2 Description

The major elements of an automatic identification system are:

1. An identifying code on the item
2. A device for reading the code automatically
3. A system for communicating the information to a decision or recording point
4. A decision or recording entity, which may be automatic or manual
5. An action mechanism to carry out the decision

These elements may vary among systems.

A schematic of a typical automatic identification system is illustrated in Figure 13.1.

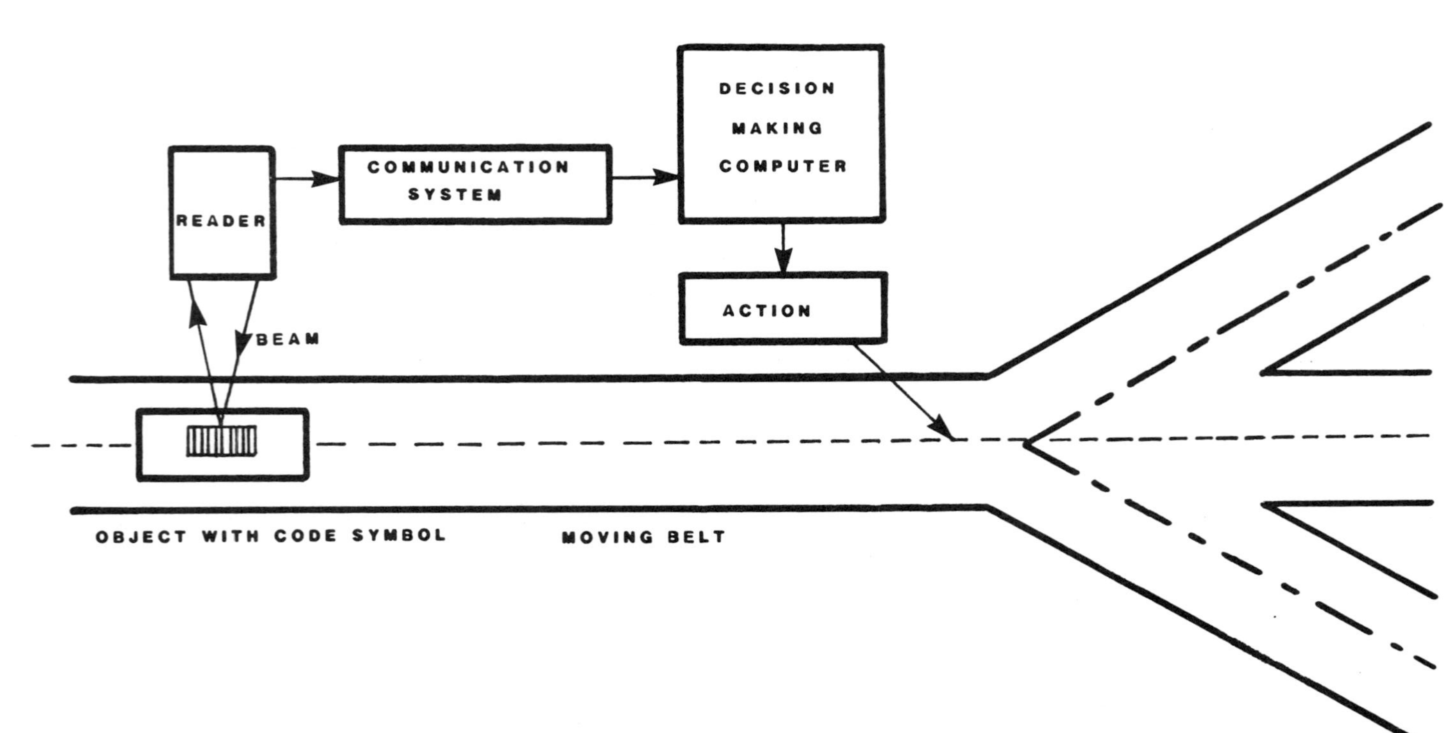

Fig. 13.1 Automatic identification system elements.

The objectives and purposes of an automatic identification system may vary from one system to another. They determine the type of system, the hardware, the software, the codes, even the printing quality. Many system parameters are determined by the customer, industry standards, transportation carrier, or other outside influence.

13.2 BAR CODES

The most common coding method is the bar code. A series of parallel lines and spaces is designed to convey the identifying data in the system. There are several different bar codes in existence, although the future of some is problematical. The commonest example is the Universal Product Code (UPC) used on consumer goods sold through supermarkets and mass retailers.

The criteria for a good bar code are:

1. Accurate reading. The code can be read accurately with minimal errors. It will be read on the first pass and not need a repeat pass.
2. Large tolerances. This allows a margin for economy and ease of printing.
3. Density. The code provides a large number of characters in a limited space.
4. Number of characters. The maximum number of characters per inch should be provided. Alphanumeric has far more characters than does simple numeric.
5. Self-checking. The code should have redundancy or check numbers which will detect single errors in coding. Two matching errors cannot be detected by normal self-checking, but are extremely unlikely to occur.
6. Widely used. The same bar code should be used by all interfacing organizations.
7. Speed. The bar code should be readable at a high rate of speed.

Many different bar codes have been developed. Some have fallen into disuse or have only limited applications. The following are the most important ones in use today:

1. Universal Product Code
2. Two of Five (2 of 5) Code
3. Code 39 (3 of 9)
4. Code 93

5. Interleaved Two of Five Code
6. CODABAR

The Universal Product Code (UPC) is widely used in retail operations. It encodes manufacturer, product, and size identification. Tolerances are tight. A code label is printed either directly onto the product or its carton or onto a separately applied label. It can be applied and used by the manufacturer, distributor, and retailer of consumer products. Functional applications include packaging, checkout, inventory and production control, and data processing.

Two of Five (2 of 5) Code was an early code used for sortation, shipping, and transportation ticket identification. It requires a larger number of bars and spaces than do some other codes. Many organizations still use it for materials handling systems.

Code 39 (3 of 9) is an alphanumeric code which permits encoding more information in higher density than does a numeric code. It is widely used in industry and has been adapted by the U.S. Department of Defense for its LOGMARS (logistics) program.

Code 93 is an adaptation of Code 39 that provides better security and density. It is widely used in industry for sortation systems, receiving and shipping, work-in-progress, data processing, and verification. It uses two check digits, which provides better security and almost completely eliminates errors.

Interleaved Two of Five Code has high density and variable length. This makes it useful for many industrial and distribution applications involving container and product verification and identification.

CODABAR is a specialty code which is versatile and has good internal self-checking features. It has been adopted for blood banks, libraries, inventory control, photofinishing, and other specialized applications.

13.3 READERS

The second element of an automatic identification system is the reader which reads the code on the item and translates this into impulses sent into the system.

13.3.1 Optical Readers

Optical readers consist of photoelectric, laser, and other beams which are interrupted as the code passes through the field of vision of the beam. These interruptions are converted to impulses sent to the system controller. The controller interprets the code provided by the impulses and issues instructions accordingly.

Fixed-Beam Reader

One of the simplest optical readers is a fixed photoelectric beam which scans the code identification on the product and reads the information. The code may be incorporated into a reflective tape affixed to the product or into a code plate or card on the materials handling carrier. The code may be printed directly onto the product or its container.

In some systems, several separate photoelectric scanners are used to identify the product from the reflective tape strips or code plates. A single fixed-beam reader services all the scanners and transmits the data as direct output to the system controller. This configuration is known as a parallel fixed-beam code reader. The parallel fixed-beam system can use simple markers such as uncoded reflective tapes. Various combinations of coding tapes provide a number of message alternatives. This is useful where the system is sorting product for a limited number of destinations. The system may use a separate scanner at each diversion fork. Although the system has several advantages, a careful cost analysis should be made of the control reader system versus the cost of separate individual readers at each junction point.

Serial Reader

A serial code reader reads a code permanently attached to a tote box, pallet, container, or other carrier. It uses two photoelectric scanners in serial. The information is transferred to the system controller, which then directs the container to the correct spur based on preprogrammed instructions.

Moving-Beam Reader

A moving-beam scanner employs a laser light source and a rotating many-face mirror. The mirror rotates at high speed, producing beam sweeps of up to 600 separate scans per second. The system is similar to that employed by physicists to measure the speed of light. It is widely used in manufacturing to identify automatically work-in-process items during processing. It is more versatile than other readers in the size of the code and the variety of inks, labels, and substrates. Label placement is not critical. The code can be printed twice at right angles to improve the readability where code orientation varies. A moving-beam reader is shown in Figure 13.2.

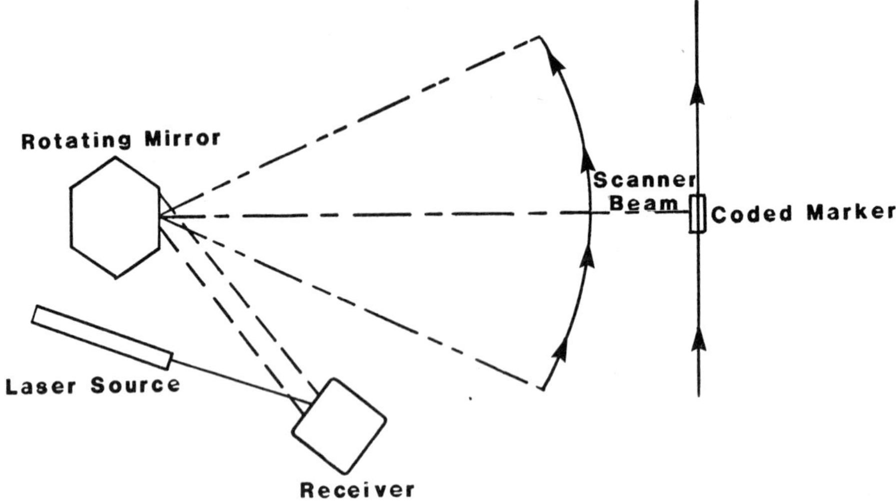

Fig. 13.2 Moving-beam reader.

Handheld Reader

A handheld reader, as the name implies, is a code reader that is held and moved by the worker. An example is the device that reads a borrower's library card to verify the borrower's status before books can be checked out. It performs the same functions as a fixed-beam reader and is used where fixed-beam or other automatic scanners are infeasible or inefficient. This includes conveyorized products of varying shapes and sizes. One can be used as a backup for an automated system or as the primary reader in lieu of more expensive or less efficient readers.

Omnidirectional Reader

The omnidirectional scanner takes many different readings of the label as it passes through the field of vision. A computer compares these readings to a preprogrammed memory of patterns, identifies the correct identification of the item, and issues the destination instructions for sortation. Because of its high cost, its use is limited to applications in which other reading techniques or equipment are unsuitable.

13.3.2 Voice

In a voice system, the operator orally reads the code information into the system. This is useful where the packages must be manually handled. A headset mike enables the operator to call the information at the same time he or she is handling the package, affixing a label, or performing some other function.

A major difficulty with a voice system is that human voice characteristics vary considerably among workers. As a result, the system may have difficulty recognizing and reading the voice signals correctly. The voice characteristics of an individual may be inconsistent due to illness, motivation, or fatigue. Ongoing research and development is continuously improving voice coding systems.

13.3.3 Radio and Microwave

In some environments, optical identification methods will not work satisfactorily. Optical methods require line-of-sight vision and absence of pollution and competing light sources which interfere with optical transmission. Radio frequency and microwave can be used effectively in such circumstances.

The three major components of a radio-microwave system are:

1. A transponder for storing and transmitting the code information. It is based on a printed circuit board.
2. An antenna near the product line to transmit signals to the tag on the product or carrier and to receive the response signal for transmission to the reader.
3. A reader that decodes and processes the data for transmission to the system controller.

13.4 SORTATION CONTROL

A key function of many material handling systems is classifying product into catagories and then distributing items according to catagory. For example, a distribution system may sort the outgoing product by destination. First, a code label with the destination and other product data is placed on the item or container. A reader picks up the destination code information as the item moves down the line. This information is transmitted to the system controller, which then directs the system to move the item to the proper destination spur or collection point.

13.4.1 Manual

Manual sortation is still used in many situations. The worker reads the destination from the label and directs the item to the correct distribution spur. This method is labor intensive and subject to error. During high-pressure rush periods, the worker may misread the label in an effort to keep product moving.

A common modification of the automated sortation system is a mixture of automatic and manual. One grocery warehouse has nine sortation output lines in its system. Eight are automated, with reading and sorting done automatically according to destination. The ninth line receives all outgoing product that does not fit into one of the eight automated classifications. Product may go to the ninth line because it doesn't fit into one of the eight classes, because it was improperly labeled, or because it received no label at all. The product on the ninth line is handled manually.

13.4.2 Direct Diverter Control

In this technology, the destination code is placed directly on the package or item. The information is read directly and the sortation action made without intervention of a human worker or computer.

One example of this technology requires a code plate attached to each item moving down the line. This code plate designates a destination. It can be permanently attached to the item, as with a tote pan, pallet, or container, or it can be affixed individually to the item at the origination point. At each diversion point, a photoelectric scanner reads the code plate and a microprocessor or other logic device guides the product to the proper destination.

Another application uses reflective tape at various heights to code the destination. Photoelectric scanners at each of the heights read the presence or absence of the reflective tape. If the photoelectric scanner reads a reflective tape at its height, it operates a diverter to guide the package to the corresponding destination.

13.4.3 Centralized Control

In this system, a code label includes destination information. The scanner transmits this information to a central system controller which accepts the data and directs the package to the appropriate destination collection point. The sortation itself is accomplished through computer-controlled diverters. This system permits tracking the individual packages through the system and controlling the flow to avoid congestion and delays.

13.4.4 Coded Belts

An older technology used magnetically coded belts. The destination is encoded at the origin and read as the product moves along the line. When a reader finds a code corresponding to its destination, the product is diverted. This method involves costly maintenance and troubles with environmental contamination.

13.4.5 Mechanical Sortation

Some older methods used various mechanical devices for sortation. An example is the steel-ball-in-groove method in which a steel ball is inserted in a groove or slot identifying the proper destination. At the diversion point, the ball makes a contact which then operates the diverter. Mechanical devices have been largely superseded by electronic devices, which are cheaper, less subject to error, and easier to maintain.

13.4.6 Microcomputer

Most modern sortation systems depend on microcomputer controls since they are more efficient and cheaper. Furthermore, a microcomputer controller can interface with other systems and can provide tracking data not available with other control systems.

In a microprocessor sortation control system, the following steps occur:

1. The package is placed on the moving conveyor.
2. A sensor detects the package and stops it when it reaches the metering point.
3. The operator punches in the destination address and the package identification.
4. The system controller receives the information from the operator's terminal.
5. The package moves on down the line.
6. Tracking data on the package are read by sensors and communicated to the system controller.
7. At the appropriate diversion point, the system controller directs the package onto the correct distribution line.
8. A completion report is sent back to the system controller.

The system can have routines to handle congestion and full-lane conditions. The package may be recirculated or diverted to an alternate spur.

The microprocessor system can have additional features. One is a fault detection routine for locating nonperforming components. Another is interfacing with other systems such as overall inventory monitoring and updating. The microprocessor system is easy to modify when the system configuration is altered.

13.5 SORTATION EQUIPMENT

There are many devices to accomplish physical sortation in an automated system. The basic mechanism is a moving conveyor. Several types are discussed in Chapter 4. The sortation device diverts the item to a spur or subsidiary conveyor chosen by the system controller on the basis of preprogrammed instructions. This is normally a rather simple mechanical device. The sophistication lies in the automatic controls.

13.5.1 Deflectors

A deflector is a movable barrier that diverts the package from the main line to a spur or sorting line. Deflectors are inexpensive and can handle up to 35–40 cartons per minute. The speeds and weights involved do not cause much wear since the deflector arm doesn't come into direct contact with the moving belt. The package must have sufficient strength to avoid damage to the product on contact with the deflector arm.

13.5.2 Pushers

A pusher is a device that pushes the package off the conveyor onto the distribution spur. It is relatively inexpensive and is operated pneumatically or hydraulically.

13.5.3 Pullers

This device pulls the package from the main conveyor onto the distribution conveyor or spur. It permits closer spacing of spurs and produces good throughput rates.

13.5.4 Paddles

Rotating paddles can be used for sortation. They are expensive to acquire and maintain but tend to be easier on the load than some other devices.

13.5.5 Pop-Up Devices

There are several devices that pop up on signal to move the packages off the main line to a sorting spur. These include rollers, wheels, belts, and chains. Each can be used in specific applications, depending on the nature of the product and the throughput rates required.

13.6 SORTATION MACHINES

A sortation machine is designed and dedicated to the sortation function. It has a conveyor system with high throughput rates and is computer controlled through a programmable controller, minicomputer, or microcomputer. It is equipped with sensors for gathering data and monitoring the action.

13.6.1 Tilt Trays

A tilt-tray system is based on a moving conveyor on which are mounted a series of trays that can be tilted either way to discharge the product. Each package is dumped onto the appropriate destination chute on which it moves by gravity to the end of the chute. When the sortation is finished (for example, at the end of the working shift), the packages are loaded onto a truck for delivery to the final destination. Throughput can be high, up to 300 packages a minute. This rate requires several stations since the maximum rate is five units a second.

13.6.2 Slat Sorters

This type of sorting machine uses a series of slats mounted on the moving conveyor. The machine can handle packages of widely varying size and weight. If the product covers more than one slat, all slats holding the item can be tilted simultaneously. On the other hand, a slat sorting machine is less flexible in configuration than the tilt-tray type.

14

Automated Storage and Retrieval Systems (AS/RS)

14.1 INTRODUCTION

14.1.1 Definition

In the broadest sense, an automatic storage and retrieval system is one that moves incoming material into storage and extracts material from storage automatically without direct handling by a human worker. This definition covers a wide variety of systems of varying degrees of complexity and size. However, the term automatic storage and retrieval system (AS/RS) has come to mean a single type of system which uses multitiered racks, a vehicle that can store and retrieve a load from any rack position in its aisle, and a computer control and communication system. The central feature is the computer, which directs, controls, and records the movement with little or no human intervention. This unit load is standardized in dimensions and physical specifications although not in content.

14.1.2 Description

The AS/RS vehicle usually runs on tracks in the aisle, although some systems use signals from a wire embedded in the floor to guide the vehicle and hold it on course. The height may vary up to 100 ft, although this height is rare. It is equipped with a movable platform which is raised to the rack location position and then lowered to floor or discharge level. Power is supplied by batteries since internal combustion engines are impractical and hot electric wire systems too dangerous.

Important characteristics of the vehicle are the velocity and acceleration. The latter controls the vehicle since there is seldom room to attain full speed. Vertical speeds of 70–80 ft/min and horizontal speeds up to 500 ft/min are possible, although these speeds are seldom attained because of system size limits. Acceleration can be as much as 1.75 ft/sec/sec, although the practical rate attainable is only a little over 1 ft/sec/sec.

The operating design limit for an AS/RS vehicle is about 30 dual-design command executions per hour. A dual command is one which requires the machine to execute two separate tasks before receiving and executing the next command. It may be more useful to work from an operating design rate of 50–60 tasks per hour.

Racks for an AS/RS are similar to those used in manual systems. However, the tolerances and maintenance are tighter since the vehicle must position itself exactly to insert and/or remove a load without damage or interference. Especially important is a firm building foundation to avoid settling, which would throw the dimensions off. Typical racks are shown in Figure 14.1.

14.1.3 Criteria for Use

Although an AS/RS can produce substantial cost savings along with increases in throughput and efficiency, it is not suited for every inventory application. Among the criteria for a successful AS/RS installation are:

1. A product line that can be effectively handled in unit loads.
2. A high throughput volume.
3. A suitable site. This includes enough area, soil conditions that will support the structure, utilities, and good access.
4. Enough inventory quantitative and compositional stability to justify an investment writeoff over a reasonable period of years and to avoid major expansive alterations to the system.

14.1.4 Examples

Many AS/RSs are in distribution centers where a manufacturer keeps his finished goods until they are shipped to the customer. The inventory may come from different sections of a large factory or from factories in several different locations. A consolidated finished-goods warehouse operated with an AS/RS may be more efficient than separate inventories at each manufacturing location. The expense of storing and shipping from a number of separate locations often exceeds the costs of inventory consolidation.

One of the original AS/RS installations in the United States is the Hallmark warehouse in Liberty, Missouri, a suburb of Kansas City.

Fig. 14.1 Racks for a rack-supported building frame for an AS/RS unit load system. (Courtesy Warehouse Storage Systems Co.)

Greeting cards, stationery, and other Hallmark products are stored in lots as they come from the factory. When called for, an item is retrieved by the AS/RS vehicle and brought to the pick station. After the pick, the remaining product is returned to a rack position. A tilt-tray conveyor is used to sort the outgoing orders by shipment destination.

Another application of AS/RS in work-in-process inventory. Partially completed work is stored by the AS/RS until it is needed. In addition to the savings in costs, especially labor, and the improvement ment in turnaround time, an AS/RS usually reduces "lost" inventory to almost zero. "Lost" inventory is product on hand that is worthless because its location is not known. An engineer with a large automotive manufacturer commented to the writer that this was a major problem in its manufacturing operations. Some systems will handle partially completed as well as finished components. An example is automobile engines. Combined with automatic identification capabilities, each engine can be tracked through the system so that its status and location are known at all times.

Another installation is an aircraft engine rebuilding overhaul facility ultimately used by the U.S. Air Force (1). After a specified number of operating hours, an aircraft engine must be overhauled. This requires disassembling the engine, inspecting each part, discarding rejected parts, remanufacturing parts that can be made reusable, and reassembling satisfactory components into a complete engine.

The system has the following characteristics and capabilities:

1. A family of container sizes to accommodate the variety of parts involved.
2. Scannable labels on each container.
3. Robots to move parts. This requires precise positioning.
4. A variety of paths through storage and manufacturing.
5. Integration of AS/RS with the manufacturing, inspection, and production operations.

14.2 DESIGN FACTORS

This section covers some of the important design features of an AS/RS. Not all factors can be considered in detail. We go on the assumption that almost every AS/RS is built from standard components manufactured to standard engineering specifications. From the user's point of view, the major design problem is selecting the right combination of components to accomplish his materials handling objectives. He should verify that the designer and manufacturer are competent and that the components and the system will perform satisfactorily.

14.2.1 The Unit Load

Basic to the AS/RS design is the unit load. If the company already has a standard unit load design in use, the AS/RS design can proceed rapidly. If the company does not have a standard unit load specification, it should examine its operations carefully to establish the best standard load design for its operations. This requires consideration of many factors, including weight, packaging, exposure to damage, number of stock keeping units (SKUs), dimensions, average turnover and throughput expected, shipping and common carrier regulations, average pick size, and customer preferences.

Dimensions

The unit load design starts with the dimensions: length, width, and height. Not only the dimensions of the overall standard unit load, but the dimensions of the individual items, must be considered. The company may have some leeway in selecting the dimensions. However, it is usually better to go with a widely used standard unit load size than to ask for a nonstandard one which will require custom-made pallets, racks, and vehicles. In AS/RS, as in other fields, custom-made equipment is inordinately expensive and seldom justified.

Length and width are usually fixed by the given standard. Height can be varied within a maximum upon which all vertical dimensions are based. However, if the average load height is much less than the maximum, cubic space utilization will be low.

Homogeneity

Will the loads be homogeneous or similar to each other in basic characteristics? If so, AS/RS design will be facilitated and its efficiency enhanced. If not, the designer will have to allow for variations in load characteristics. One solution is the standard pallet or container which will hold most of the company's different products. Another is to allow variations in some pallet and rack dimensions, although this normally adds to the expense.

No matter what the company's product, there will be items that cannot be accommodated in the AS/RS. These items must be stored off-line in a special facility. This is expensive and inefficient, though necessary. Every effort should be made to design the AS/RS to accommodate as high a proportion of the company's products as possible.

Fragility

The possibility of product damage should be considered in AS/RS design. If the product is easily damaged, the system must be designed to minimize product accidents. An example of easily damaged product is furniture, which can be easily scarred, nicked, or otherwise damaged. These damage marks reduce the item's value substantially so that the item must be sold at a large markdown. This factor has seriously inhibited the use of mechanized materials handling in the furniture industry.

Weight

The weight of the standard unit load determines several design parameters. An important one is the total rack height of the system. A heavy standard load will limit the overall height of the structure owing to the load-bearing limits of foundation soils and floors.

14.2.2 Racks

The rack structure for an AS/RS is built of standard components which will meet all load, deflection, and other engineering requirements if properly installed. The soil conditions and the foundation must be capable of bearing the load without settling. The number of racks is determined by the space available and by the systems requirements. If air space is available, height can go to 100 ft. This will increase construction cost and will require a taller stacker vehicle. The operating expense may be greater since vertical travel is slower, about one-third the speed of horizontal travel. However, the greater height may be justified economically if land cost is very high or if the facility must be fitted onto a small site.

A good rack sytem will last 20 years or more without replacement. However, changes in system requirements will usually require modification before the end of the useful physical life. Also, changes may be made in the control system and in the vehicles without reconstructing the rack system.

14.2.3 Pallets

Most AR/RSs use a pallet as the basic product-holding unit. Pallets are inexpensive and are widely used for storage and material handling. A uniform standard pallet simplifies the operations of the system and yet provides a good platform for a wide variety of products.

Standard

Many AS/RS use standard pallets such as those described in Chapter 4. This policy has the advantage that the pallets are interchangeable with those of interfacing systems using the same standard pallet. For example, product may be mounted on a pallet at the vendor's, shipped by common carrier, and moved directly to AS/RS storage without transloading. Standard pallets are cheaper and easier to obtain and maintain. For most installations, they are quite satisfactory.

Slave

Some installations use slave, or dedicated, pallets that do not leave the system. They can be custom-built for a specific purpose. For example, a pallet can be designed to hold a specific type of engine or irregularly shaped casting. On the other hand, they do require a mechanism to place the product on the pallet at the receiving station and to remove it at the outgoing station. This may be done by a robot, although other types of special-purpose mechanisms are available.

Custom

Other than a slave pallet, a company may require a nonstandard, custom-designed pallet. The system designer has considerable leeway in specifying a custom pallet, although there are some restrictions. The custom pallet should have the same dimensional base as the standard AS/RS pallet. Otherwise, the whole system becomes a custom project with exceptionally high capital investment costs. Also, the custom pallet must meet the engineering requirements of the system, such as weight, size, and dimensions. A custom pallet is seldom worthwhile unless it is a special attachment to a standard pallet as a base.

14.2.4 Structure

The structure is the framework within which the AS/RS operates. It must be solid and able to meet load, deflection, stability, and other engineering specifications.

There are two major types of AS/RS structure. Originally free-standing racks were installed within the factory building or in a separate building. A later development is the rack-supported building in which the AS/RS racks support the rest of the structure, including the roof and walls. See Figure 14.1.

The choice between the two types depends primarily on economic considerations. Either can be designed to be structurally sound. If the company already has a suitable building adequate for the AS/RS, it may elect to go with the free-standing racks. Also, if the racks are limited in height, about 50 ft or less, the free-standing rack is usually cheaper since it can be enclosed in an inexpensive preengineered building with a minimum of costly special features. Above 50 ft height, the rack-supported structure is usually less expensive. Heights above 75 or 80 ft create some difficult engineering problems in foundations, soil conditions, and vehicle heights. However, a few systems approach 100 ft in height, and more are expected as experience is gained and technology improved.

14.2.5 Environmental Considerations

Environment affects the design. Extremes of ambient temperature influence building specifications of insulation, snow loading, heating, and cooling. Some products require close temperature control. An example is a frozen- or fresh-food warehouse. Energy needs differ. The geology and soil characteristics of the site may limit the size and type of building structure. A foundation on bedrock will support a larger, heavier structure than one on unstable soils.

14.2.6 Institutional Factors

Institutional factors affect or limit the company's options in structure design or operation. Local zoning codes, airport restrictions, property boundary lines, deed restrictions, and fire codes are examples.

14.3 SYSTEM THROUGHPUT

System throughput is the number of loads moved into, through, and out of the system in a given period. It is the measure of the operating capacity of the system in contrast to the static capacity, a measure of how many loads can be stored. The throughput rate is a function of both the number of loads stored and the capabilities of the storage and retrieval vehicles. It is also affected by control routines, design objectives, type of operation (unit loads only or order filling and kitting), and system reliability. An important feature is variability. The AS/RS is designed to attain a given throughput. However, most operations have demand fluctuations during the day, which may cause delays at peak times yet have unused capacity during slack periods.

14.4 AS/RS SIMULATION

Simulation of an AS/RS is a good way to determine the operating abilities of a proposed design or the design parameters to accomplish the system's mission. The inputs to the simulation are randomly selected locations and tasks corresponding to the system design and the user requirements. The simulation model calculates the time needed to perform the tasks on either a single- or dual-command cycle. The output is a summary of operating results in number of tasks accomplished per unit time period, such as an hour.

The simulation model can also accommodate fluctuations in requirements throughout the day. It is common for a system to have more replenishments than retrievals in the morning, a near balance at midday, and a preponderance of retrievals during the last hours of the workday. This variation may require separate runs for each period that has substantially different characteristics and requirements. Stratification of requirements by period will help determine the capacity needed for peak periods and may also provide clues for easing fluctuations to reduce the peak hour capacity needed.

14.5 DESIGN FEATURES

Several different design concepts are available for an AS/RS system. These call for different combinations of the number of S/R vehicles, the number of aisles, the number and type of storage racks, and the type of S/R vehicle. As with any engineering system project, the choice of design concept calls for tradeoffs among the sytem components. For example, more aisles will require more vehicles if each vehicle is confined to one aisle. If fewer vehicles than aisles are used, provision must be made for vehicle transfer between aisles.

On the other hand, more aisles will allow a lower overall rack height and shorter vehicles. Each AS/RS is a separate design project in which concepts and tradeoffs must be evaluated individually.

14.5.1 Numbers of Aisles

The number of aisles is usually determined by the building space available. The building size depends on size of the lot, zoning and building code requirements, and construction costs as well as system requirements. The number of aisles can be reduced by increasing the rack height.

Another determinant of the number of aisles is the width of the rack and the width of aisle required for the vehicle and for clearance. System volume requirements will influence the number of aisles. Provision should be made for expansion and for modification of the system. For example, a more complex control system may reduce the number of storage spaces and thus the number of aisles.

14.5.2 Number of Machines

The most common AS/RS system design calls for one vehicle per aisle. This is simpler to design and easier to operate and maintain and has no problem of vehicles interfering with each other. If there is more than one vehicle per aisle, the control system must be much more complex to avoid interference and accidents and to operate efficiently. If there are fewer vehicles than aisles, provision must be made for transferring vehicles between aisles. This requires a transfer car to move a vehicle from one aisle to another. The usual pattern is one transfer car to each S/R vehicle. It is possible to have fewer transfer cars than S/R vehicles, but this requires an extrememly complex control system. The transfer system is suggested only when the throughput rates are low in relation to the number of loads stored, with only a few cars in relation to the number of aisles needed.

14.5.3 Double-Deep Storage

A double deep storage system has racks that accommodate two loads deep from the aisle. This is appropriate when the system has many loads of a limited number of SKUs and a low throughput rate. One approach is a single-width aisle with a shuttle mechanism capable of storing the load in the second position and retrieving it from that position. It is preferable to put two loads of the same SKU in a position to eliminate the need for shuffling loads to get at the one in back. Some systems use a double-width aisle with a machine capable of handling two loads side by side. This improves the system productivity. However, the vehicle complexity and consequently the cost increase with the length of extended movement required and with the load capacities needed. A double-deep system is better when loads are stored or retrieved two at a time.

14.5.4 Deep-Lane Storage

In some industries, such as food distribution, there are few different SKUs but a large inventory of each SKU. Deep-lane storage is efficient in this situation. The system uses a gravity rack with room for several loads on each rack. An S/R machine picks up the incoming material from a conveyor or other input system and deposits it on the high side of the assigned rack. The load moves by gravity until it contacts the last previous load. At the other end of the rack, another S/R vehicle removes the load as it is called for and transports it to the discharge station. This system ensures that product is used on a first-in, first-out basis. It also provides high-density storage and efficient throughput.

14.5.5 Pick-Tunnel Storage

A pick tunnel may be used for an inventory of few SKUs with individual order-picking capability. The incoming product is stored from an outer aisle by an S/R vehicle. The inner aisle is used for order picking that may be done manually. Some systems use a man-on-board vehicle for the order picking. Several different configurations are available to suit the needs of the system and the space and capital available.

14.6 ORDER PICKING

Order picking is retrieving individual product items from stock to fill internal or external orders. It is a major function of AS/RS. It may be viewed as breaking down wholesale inventory into retail lots to meet customer demand. Order picking is a major expense in distribution. Reducing its cost is an important goal of automated materials handling systems.

Manual order picking in a traditional shelf warehouse required too much time to be spent in unproductive walking along the aisles to get the requested items. At current levels of wages, benefits, and overhead for this labor catagory, manual order picking is prohibitively expensive. However, it is used to a limited extent for products that cannot be fitted into the AS/RS or for special situations.

14.6.1 End-of-the-Aisle System

The system may be configured to bring the material to the order picker at a fixed station at the end of the aisle. The vehicle retrieves the container from its storage location, brings it to the pick station, and places it on the pick table. The order picker selects the required parts shown by a hard-copy order or, more commonly, by a cathode-ray-tube (CRT) terminal. After the pick is completed, the operator signals the system to return the container to its location.

14.6.2 Man-on-Board System

In this configuration, the order picker rides the vehicle. The machine moves to the desired location where the picker selects the required items. The vehicle is then directed to the next product location. The AS/RS may provide direct communication with the vehicle, which can have a keyboard and CRT. The operator can read the order displayed on the CRT and pick accordingly. He can also send reports of completed picks. The system may determine and command the next move without operator intervention except for the ready signal.

The man-on-board system is suitable for light product items. Also, the shelves and product must be conveniently placed for the operator. This system permits the operator to complete one or more orders without returning to a base station for further commands. At some point, the operator must return to a discharge station to turn over completed orders.

14.6.3 Multiple Functions

The end-of-aisle configuration provides space and equipment for the order picker to perform associated operations. These include packaging for shipment, labeling, order checking and verification, preparing shipping documents,and affixing meter labels for postal charges. This is not feasible with the man-on-board system.

14.6.4 Heavy Material

Unlike the man-on-board configuration, the end-of-the-aisle pattern permits handling products too heavy for human lifting and handling. The work station can be provided with separate hoists or transfer mechanisms for heavy parts.

14.7 INPUT/OUTPUT SYSTEMS

Although an AS/RS is a self-contained entity for its own operation, provision must be made for entering product into the AS/RS and for removing it when called for or when an outgoing order is filled. It is seldom feasible or economical to use manual or fork-lift materials handling for these purposes. The designer must consider AS/RS interfaces with other systems as an integral part of AS/RS design.

14.7.1 Conveyors

Conveyors of various types are commonly used for introducing material into the AR/RS. The conveyor may be loaded in one of several ways—fork-lift truck, gravity, another conveyor, or a transfer mechanism. Normally, the product is moved on a standard pallet, although some systems use slave pallets. The pallet carries the product to the entry or pickup spur of the AS/RS, where the product is transferred to the AS/RS. Conversely, a conveyor picks up the outgoing pallets from the discharge or output spur when the product leaves the system.

The conveyor can be one of several types depending on the product and on system needs. Live-roller conveyors are popular, but require some provision to avoid jamming when congested. Gravity, unpowered rollers are cheap and useful when the configuration permits. If the conveyor must change direction, a turntable, transfer, or other device for proper orientation is necessary. Most systems incorporate a mechanism for squaring the load and orienting it for proper positioning in the AS/RS.

14.7.2 Load Measurement

Load measurement devices may be incorporated into the AS/RS, often at the interface with the incoming conveyor. Each system will include a size check to see that the load length, width, and height are within the limits of the system. The load must clear all aisle obstructions and structures and must fit into the storage space assigned. It is rarely feasible to have different limits for different parts of the system. A scale is used to check adherence to weight limitations.

Some systems incorporate an automatic weight-count station for large quantities of small items. A container of standard weight is used. A few items for the container are placed on a special sample scale. The instrumentation then computes the number of parts from the net weight (gross less container) of the parts and the weight of the sample parts.

14.8 SPECIAL DEVICES

Some industries require a special device. One example is a device in a grocery warehouse to ensure that all cartons in a load are properly placed on the pallet and that shifting has not occurred. Another special device is a stretch- or shrink-wrap station for packaging an outgoing load.

14.9 INTERFACES

Although much discussion and occasionally even design deals only with the AS/RS as a separate entity, it is part of a larger production or distribution system. As such, it must interface with other systems of various types, and the designer must provide for efficiency of those interfaces.

An example of an interface is combining a power-and-free conveyor through production processes with an AS/RS for storage. The product item on a carrier is moved through the factory on a power-and-free conveyor. The AS/RS machine removes the carrier from the conveyor and stores it with its load in its rack system. It remains on the carrier until it is called for and sent on to another power-and-free conveyor for further processing, assembly, inspection, test, or shipment.

Another example of an interfacing system is one that delivers product to an automatic guided vehicle (AGV) in a flexible manufacturing system. In this case, the AS/RS is not suited for delivering the raw material or work in process to the FMS station. The AGV has the necessary mobility, flexibility, and accuracy to service the FMS stations.

14.10 AS/RS CONTROLS

14.10.1 Introduction

The control system is the key to successful AS/RS operations. Without good controls, the system will be expensive and ineffective. Modern computers have made AS/RS technology and application possible. In this section, we will discuss some of the major features of AS/RS control systems.

14.10.2 Programmable Controllers

Modern programmable controllers (PCs) are efficient, fast, and easy-to-program small computers. The input/output capabilities and memory have greatly increased in recent years. PCs are widely used in AS/RS to control the vehicles. Their flexibility has enabled them to replace special-purpose microprocessors, "hard-wired" systems, and on-board relay servomechanisms. See Chapter 12 for a further discussion of PCs.

14.10.3 Position Sensors

Position sensors are devices that verify the position of loads and vehicles in the AS/RS system. Current technology uses coded reflective labels on each storage rack and diode sensors that emit infrared radiation on the vehicles. The beams are not visible to humans and are not affected by environmental pollution or by other radiation sources.

14.10.4 Full/Empty Bin Detectors

The AS/RS vehicle must be able to detect whether the assigned bin is occupied or not. Otherwise, it will carry out a command to insert a load into a space even though the space is already occupied. Obviously, if this occurs, something has to give way with resulting damage to the product, the racks, or the vehicle.

14.10.5 Control System Reliability

An essential feature of a successful AS/RS installation is high reliability. This is defined as the percent of available time that the system is in operation. Without reliability, the idle time will rise to uneconomical levels. It is normally not feasible to have manual backup due to the configuration of the system and the high labor cost of stand-by personnel.

Increased reliability has come from several developments:

1. Better contact limit switches to replace proximity switches
2. Better scanners for automatic identification
3. Redundant circuitry
4. Better photoelectric devices
5. Packaging and pallet improvements
6. More attention to design for reliability
7. Better quality control of component manufacture
8. Better-trained and more experienced personnal

14.10.6 Data Collection

Unlike early AS/RS, present systems can provide two-way communication to the AS/R machines at any time. Rapid communication is now possible even when the vehicle is moving. This is especially important in data collection. The system controller can receive real-time reports on moves, locations, vehicle status, and inventory ins and outs. The system controller can update its database in real time and plan its operations accordingly. A vehicle need not return to a base point to receive new commands. It can be commanded to go directly from its last function point to the next assignment.

REFERENCE

1. E. J. Budill, "Integration of Automated Handling and Flexible Manufacturing," Proceedings, 1984 MHI Material Handling Forum, Houston, TX (March 27, 1984).

15

Miniload Storage Systems

15.1 INTRODUCTION

15.1.1 Definition and Description

A miniload storage system is a small-scale automated storage and retrieval system (AS/RS). The difference between the two lies in the size of the physical system, the number of items handled, the size of those items, and some operational features. The characteristics distinguishing AS/RSs from other systems are the same no matter whether the AS/RS is miniload or full scale.

Among the features common to all AS/RSs are:

1. Automatic removal of the container or pallet from its storage space, bringing it to a designated point for processing, and automatically returning the container to an assigned storage space
2. Operating control through system controllers, microprocessors, and programmable computers and through a software system designed specifically for the AS/RS functioning

The primary reasons for choosing a miniload storage and retrieval system are:

1. Enhanced throughput
2. Space limitations and restrictions
3. Smaller investment

4. Volume too small for a full-scale AS/RS and too large for a manual system

15.1.2 Examples

Among the practical miniload AS/RS are the following examples:

1. A distributor of small fasteners switched to a miniload system for finished goods. The fasteners are packaged in small cartons which fit efficiently into the miniload carriers. Storage space and turnaround time were cut substantially.
2. A common miniload application is maintenance spare parts and supplies. Most items fit easily into miniload containers. The maintenance inventory can be kept separate from production inventory. Waiting time by expensive craftsmen is reduced.
3. Another popular miniload application is assembly parts storage for small items. One electronics company saved about 50% in inventory carrying and operating costs over the costs of the previous shelf system. The company reports substantial reduction in inventories due to increased certainty and speed of retrieval of needed parts.

15.2 OPERATIONS OF A MINILOAD STORAGE SYSTEM

15.2.1 Commands

In a miniload system, the parts bin is selected by the operation controller (or the operator), retrieved by the aisle stacker or vehicle from its position, and brought to the operator work station. The operator performs his function, such as picking orders or replenishing parts. When the operator is finished with the parts bin, he initiates the control command to the stacker to return the parts bin to its assigned place. The next parts bin is called and the process repeated.

There are several physical alternatives and computer control techniques that improve the productivity of the system. These include queuing parts bins and alternating work stations. These are discussed later in the chapter.

15.2.2 Bin Retrieval and Delivery

The major cost savings in any AS/RS come from elimination of operator walking to and from the storage location. Instead, the system automatically goes after the bin and delivers it to the operator's work station. This greatly increases operator productivity.

After the stacker vehicle receives the command to retrieve a specific bin, it moves quickly along the aisle to the bin location.

Horizontal and vertical movement are coordinated so the vehicle arrives at the correct station with minimum time and motion. The system mechanism spots the platform within tolerances necessary for smooth extraction of the bin. After positioning the platform, the extractor moves horizontally at right angles to the aisle, extracts the bin from its position, and centers it on the stacker platform. The stacker then moves to the assigned work station, where the extractor reverses the previous procedure and moves the bin onto the work station platform. The stacker is then available for the next trip.

15.2.3 Operator Functions

A miniload AS/RS can be designed for various patterns of operator functions. The usual functions are:

1. Order picking. The operator picks items from the bin to fill an order for internal or external delivery. He receives this order either in hard copy or through a cathode-ray-tube (CRT) display. In some systems, he rechecks the order after completion to ensure that it has been filled properly.
2. Replenishment. The operator restocks the depleted bins to maintain the inventory at the target level.
3. Kitting. The operator picks the parts to make up a standard or special kit. It may be a standard set of spare parts, a set of parts necessary for a retrofit, or the bill of material to complete one assembly. Some systems allow the operator to prepare several kits at one time.

The operator may perform other associated functions such as:

1. Label preparation. He may prepare parts, assembly, or shipping labels. Usually, this is done through the computer system using a local printer.
2. Shipping document preparation. The system may provide for preparing documents for shipping and billing. This is difficult to do ahead of time as there may be shortages and backorders. Again, this is usually done by the computer system using a local printer.
3. Inspection. A few systems call for the operator to inspect items. This is rare except for visual inspection for obvious defects.

15.2.4 Bin Return

After the operator is through with the bin he is working on, he activates the return of the bin to its position. This command operates only on the operator's signal. Otherwise, the return could start before the operator is through or delay after he has finished. Bin return is the reverse of bin retrieval. The stacker loads the bin on its platform and moves to the assigned location, where the extractor discharges the bin into the rack. The stacker is then available for the next assignment.

In a miniload AS/RS, the bin is usually returned to its original location. However, if the system has random-access capability, the bin may be stored in the most convenient location. This requires that the system keep track of the location of each bin so the selected bin can be retrieved correctly the next time it is called.

15.2.5 Record Keeping

The miniload system may provide for keeping certain company records, such as inventory status. This requires that issues and receipts be entered as they occur. The entries can be made in conjunction with another operation such as preparing the shipping ticket and the receiving report. It is seldom advisable to ask the operator to enter such data separately. His attention is on getting the orders picked and not on record keeping. Good records are essentail to the company but are not perceived by the operator as vital to his performance evaluation.

15.3 COMPONENTS OF THE SYSTEM

In this section, we'll discuss briefly some of the component parameters that enter into the design of a miniload AS/RS. Most system purchasers will use standard components. Custom-designed and built parts are rarely cost justifiable and are seldom sufficiently better than standard components to warrant the extra design and procurement expense, not to mention delivery delays common when custom items are ordered.

Several manufacturers will design,make, and install a turnkey miniload AS/RS. This may be the best route since it centralizes responsibility and avoids the difficulties of dealing with different vendors and contractors on the same project. However, the company's requirements may dictate using more than one vendor. In this case, the company may select a general contractor who will have overall responsibility or appoint a project manager from within the company to oversee procurement and installation and to coordinate outside contractors, vendors, and company personnel.

15.3.1 Bins

The bin is the basic storage unit of a miniload AS/RS. It is smaller than the typical pallet-mounted bin. It may be made of steel, aluminum, or tough plastic. Bins come in standard sizes. Other parts of the system, such as the racks and stackers, are dimensioned to handle standard-sized bins. An off-size bin normally requires that other parts of the system be custom designed and built.

The standard-size bin may be modified to suit the company's application. One modification is a set of special racks or holders within the bin for irregular or easily damaged parts. Another is a set of partitions to separate different items stored within one bin. This separation may also be accomplished by storing the parts in original containers in the bin until called for.

Number of Bins

There are two methods to determining the number of bins needed. One is to analyze the number and characteristics of the items to be stored and the inventory quantities of each required. From these data, the number of bins can be calculated. This approach may require modification if the space required for the calculated number of bins is too expensive or not available.

The other method is to determine the amount of space available and calculate the number of bins that can be fitted into this space. This is often used when the miniload system must be fitted into an available space. It may require that the inventory to be handled be reduced in range or quantity or that the inventory policy be changed to handle some items outside the miniload system.

During this and other design phases, attention to future trends and possibilities should be considered. If the volume handled is expected to increase, provision must be made for expansion. If the nature of the inventory may change radically through a change in product line, the designer must allow for major modifications of his system in the future.

15.3.2 Stackers

The stacker has three major parts, the column, the moving platform, and the extraction mechanism. It is mounted on a vehicle that moves horizontally along the aisles, usually on tracks. The platform moves vertically up and down the column, and the extractor moves horizontally from the column to extract and replace the parts bins.

The column and platform move simultaneously along the track at speeds up to 250 ft/min horizontally and 90 ft/min vertically. Acceleration and deceleration are controlled to provide smooth stopping at the correct position and movement at full speed during the central part of the move. The system controls stopping location to an acceptable tolerance at the bin position and the work station.

One way in which miniload systems improve floor space productivity is through elimination of manual access aisles. Since all retrieval and storage is performed by the miniload vehicle, there is no need for aisle space for a manual worker. The only aisle space needed is for maintenance and fire protection. In addition to better space usage, elimination of manual access reduces the opportunity for pilferage and the amount of "lost" inventory.

15.3.3 Space

A factor in the economy and efficiency of any materials handling system is the space it requires. A major advantage of an automated materials handling system, including a miniload type, is the greater thoroughput per unit of space. This is important not only from the cost point of view, but because available space may be limited and because spreading materials handling and storage over a greater area increases travel time with corresponding decrease in productivity.

A miniload system rarely rises to the heights possible with a full-scale AS/RS. The smaller bin size, the nature of the operation, the smaller unit of pick, and other factors limit the practical size of a miniload system in both area and height.

15.3.4 Throughput

The throughput, or number of transactions per unit time, is a key parameter of a miniload system. It has two components; the operator time and the travel time of the stacker. The throughput is determined by the characteristics and capabilities of the miniload system.

There are ways to improve the thoroughput of a miniload system. Unfortunately, each involves extra investment and often extra operating expense. Whether the extra cost is justified by increased productivity and throughput must be determined by analyzing each case individually.

Ways of increasing throughput are based on either minimizing delays by the operator or the stacker or having all system components operating simultaneously to the extent feasible. Most are obtained by modifying the system configuration at the operator end. The next section discusses various configurations.

15.4 CONFIGURATIONS

Several different miniload system configurations are available. The choice depends on the user's requirements, the space available, and the capital cost compared to the benefits obtainable. The objective is to improve the operator-machine interaction to increase productivity.

15.4.1 End-of-the-Aisle Work Station

In this configuration, the stacker delivers the bin to a station at the end of its aisle, usually through a window to a platform beyond the storage rack limit. The bin stays on the platform while the operator picks from the bin or replenishes it. This layout is shown in Figure 15.1.

The end-of-the-aisle configuration is the least expensive. However, it is relatively inefficient since the stacker is immobile while the operator is working the bin. Also, the operator cannot perform pick or restock while the stacker is returning the bin just worked and retrieving the next one. These times are normally too short for the operator to perform any other useful work such as labeling or wrapping. If the dwell time is short, system requirements are not rigorous, and capital is short, this system may be the best. In addition the other disadvantages, this configuration permits only one worker to pick parts at any one time.

15.4.2 Side-Shuttle Work Station

In this configuration, there are two side-shuttle stations at the work station end of the aisle. As the bin arrives at the end of the aisle, it is shuttled to either the left or right work station, depending on which is vacant. The operator can work one bin while the stacker is returning the completed bin to its position and getting the next bin to be worked. The new bin is then moved onto the unoccupied work station. Thus, the operator and the stacker can keep working while the other component is functioning. This is a considerable improvement in productivity over the end-of-the-aisle configuration. This is diagrammed in Figure 15.2.

This configuration works best when the operator time equals the stacker storage and retrieval time. However, time must be allowed for moving the bins on and off the side positions. If the dwell times are substantially different from the stacker cycle times, imbalance and delays will result.

The system does permit two operators to work simultaneously, one for each side station. However, this reduces the operator productivity to that of the end-of-the-aisle configuration.

15.4.3 Dual-Point Conveyor

This configuration provides space for queuing bins on two work stations. The layout is shown in Figure 15.3 When the operator has finished a bin, the stacker removes it and returns it to the assigned position. This configuration permits two workers to work

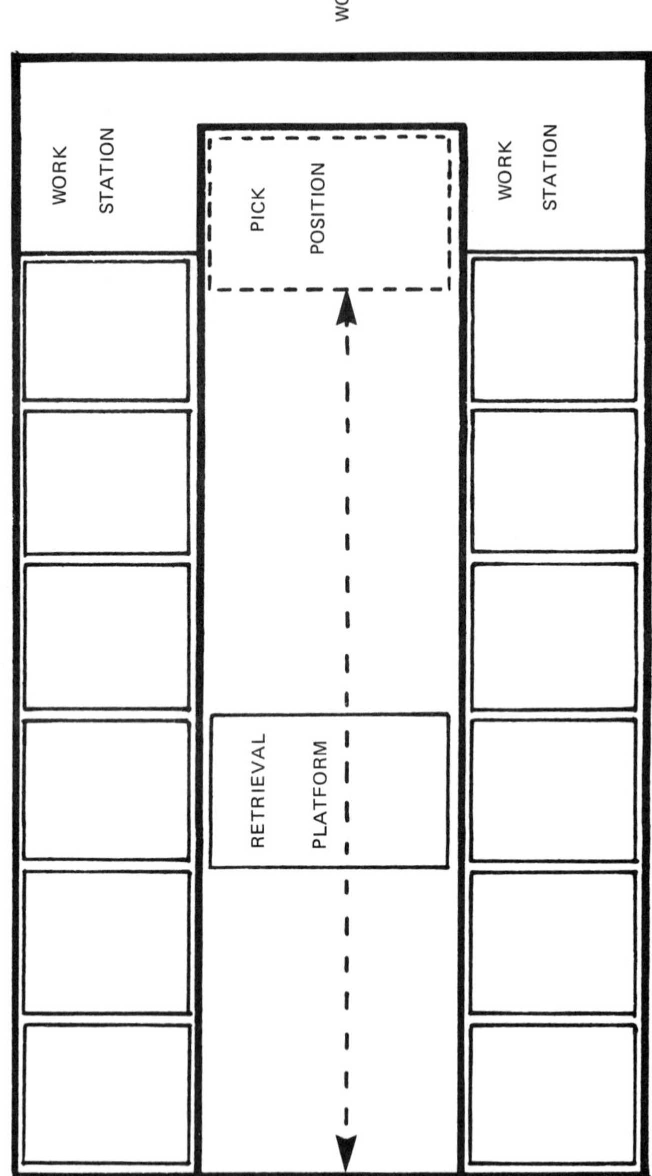

Fig. 15.1 End-of-the-aisle work station.

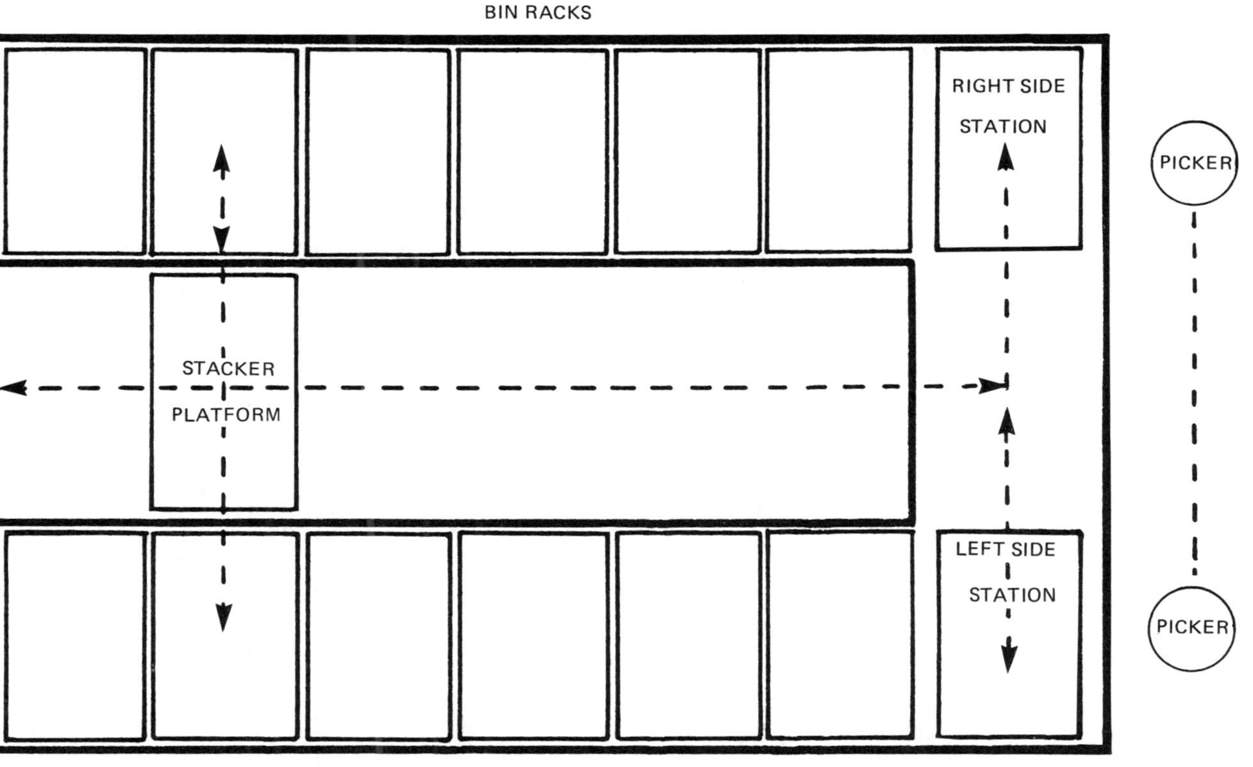

Fig. 15.2 Side-shuttle work station.

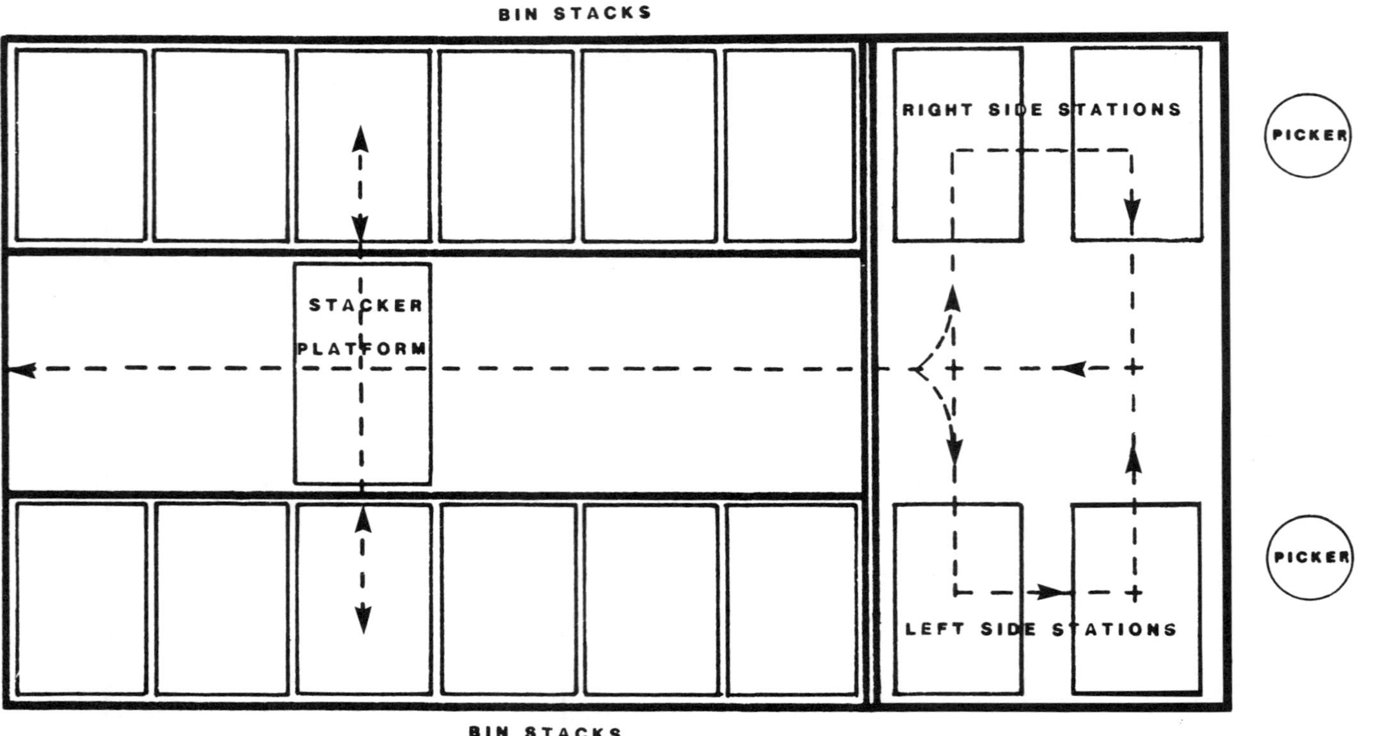

Fig. 15.3 Dual-point conveyor.

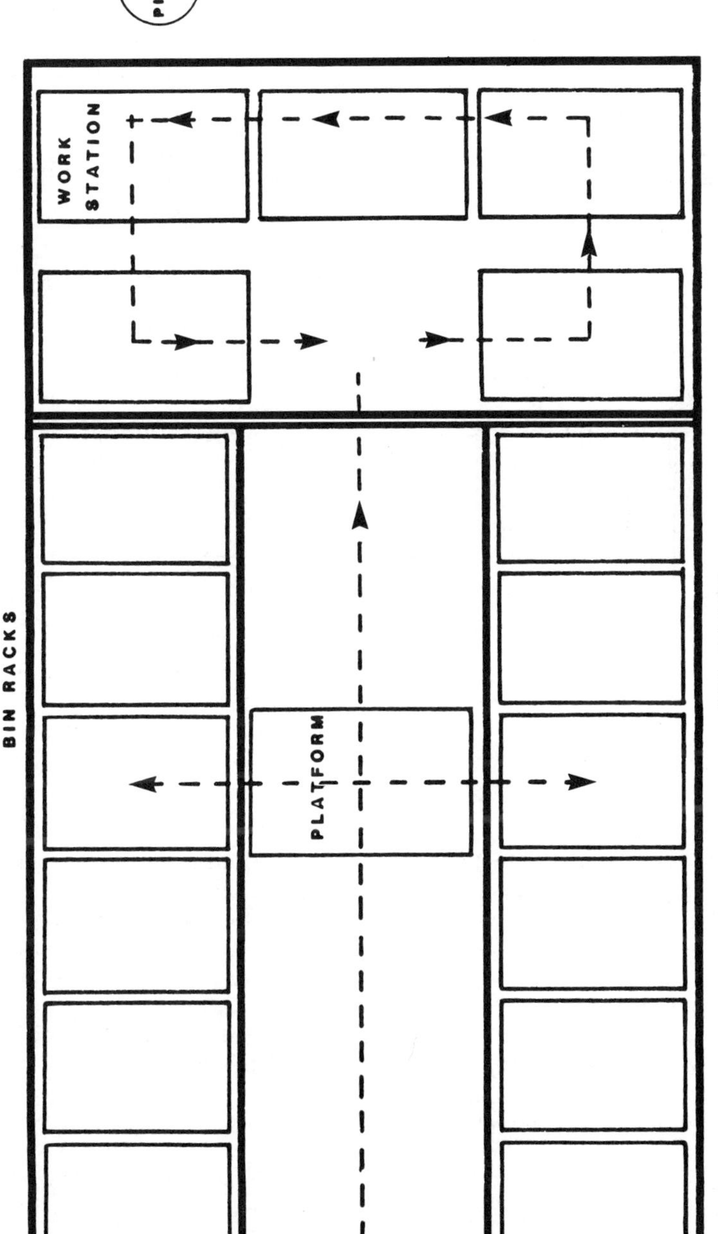

Fig. 15.4 Horseshoe conveyor.

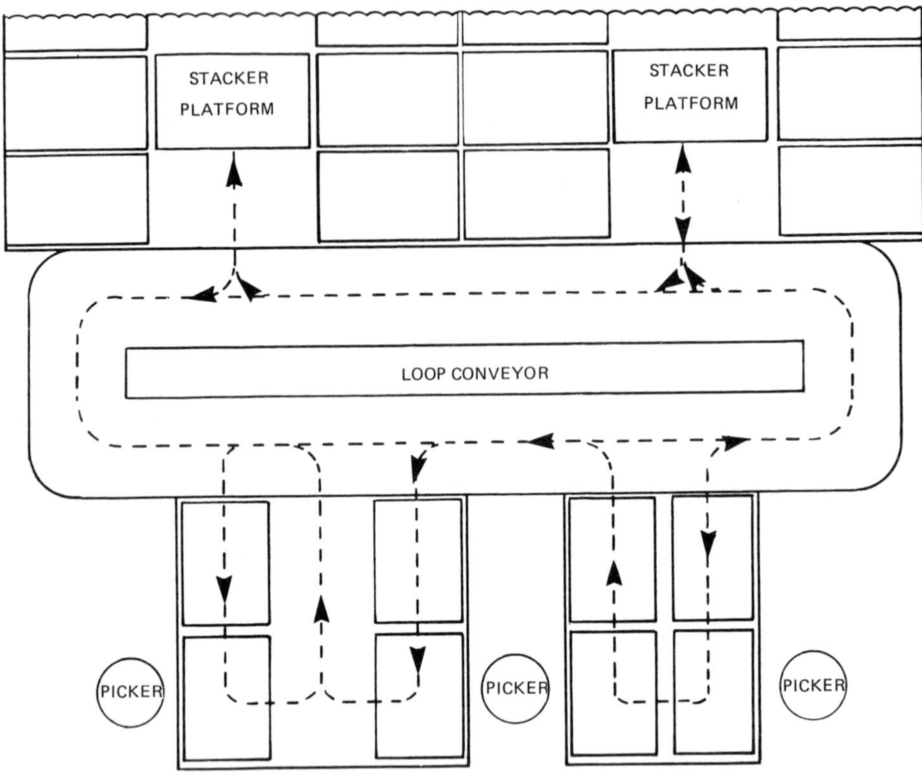

Fig. 15.5 Loop conveyor.

separately and independently. The configuration improves overall productivity of the system and the operators.

The major disadvantage of this configuration is inflexibility. A complete bin must be removed immediately, or the operator must wait. If the stacker is busy when the bin is completed, the operator must wait until the stacker is free to remove the completed bin. Once the queue is in motion, the order cannot be changed to balance the workloads or to move a priority pick to the head of the queue.

15.4.4 Hourseshoe Conveyor

In this configuration, the operators work independently of the stacker. They don't have to wait for the stacker to bring or remove a bin.

This layout is shown in Figure 15.4. The conveyor provides a buffer in the bin flow to and from the work station.

This system may be subject to delays if the operator work time on a bin is greater than average. This causes delays to other operators, which reduces the efficiency of the entire system.

15.4.5 Loop Conveyor

In this configuration stackers move bins onto and off the loop conveyor without reference to the operator activity at the time. (See Figure 15.5.) Several operators can work simultaneously. Each has access to bins. Some locations can be devoted to picking and others to replenishment. These functions can be changed during the day according to the volume of orders being processed.

The loop configuration is useful in kitting. The operator can pull a quantity of parts from each bin to make up several identical kits at a time. Or one kit can be made up at a time by pulling parts serially from the bins. There is plenty of buffer capacity so both the operators and stackers are not held up by delays of other components of the system.

15.5 COMPUTER CONTROLS AND MICROPROCESSORS

15.5.1 Introduction

In common with full-scale AR/RS and other sophisticated materials handling systems, computer controls and microprocessors are essential to the feasibility and success of a miniload AS/RS. Without computer controls, a system would have to be operated manually. This mode would eliminate many of the automated functions now possible with computers. We would be back to the expensive, slow, labor-intensive, error-prone manual methods that have become obsolete for mass, high-volume operations.

15.5.2 Entry Commands

Commands can be entered in one of several ways:

1. Manually
2. On punched cards
3. Operator entry to a computer terminal
4. System controller direction with possible operator override or supplement

Manual System

In this mode, the operator initiates all necessary commands. He enters the calling information for the bin or part he wants into the system. When he is finished, he commands the bin removal and return to the assigned location. This method is least costly, but is also the slowest and most subject to delays and errors. It may be quite satisfactory if the system throughput doesn't justify a more sophisticated, expensive system. This would be true if the system is used primarily for tool cribs where the work is intermittent and the throughput requirements are low.

Punched Cards

Many older systems relied on punched cards for data entry. The card contained information on bin location, bin destination, and bin contents. It could also contain data for ticket printing if this function is included in the system. From the card information, the card reader activates the stacker through the stacker microprocessor. This system relieved the operator of several functions. However, it has been superseded for most installations by on-site computer terminals.

Operator Computer/Terminal Control

In this system, the operator is provided with a CRT and a keyboard terminal. Pertinent information such as location and inventory status can be called up and displayed on the CRT. The system takes over and carries out the commands. In some systems, the command is processed and edited. For example, if several bins are called up at one time, the computer may compute the most efficient sequence and issue commands accordingly.

This system can also be programmed to collect and store data. For example, issue quantity and description can be collected and entered into the perpetual inventory. Bin location can be monitored if changes are made. A further function is converting issue data into accounting charges for job order and cost control.

System Controller

A further level of development is the system controller, which receives the order-picking and replenishment information directly and issues commands to the system. It may provide for operator interaction through override or by leaving certain commands to the operator. For example, the operator may signal release and return the bin to position after he is through. Otherwise, the system might take his bin away before he is through with it.

The computer can perform other functions. Terminals or readouts can be installed at each work station. These can provide information to the operator on items to be pulled, the quantities, bin location, and order designation. The digital readout display has the advantage that information used by the operator must be displayed. This simplifies his task and reduces the possibility of errors.

15.5 3 Sequencing

Sequencing is arranging system components in the most efficient order for the task to be accomplished. For example, the computer can examine the bin locations of the parts called for on an order and arrange the sequence of calls to minimize the distance and time of the stacker travel. The time to perform sequencing manually would be excessive, but by computer it is so short that it has no noticeable impact on computer operations.

15.5.4 Queuing (or Buffering)

Queuing is arranging a series of objects to be serviced into an organized waiting line. In the miniload system, it is arranging a series of bins waiting to be worked into a line. Normally, bins are worked in first-come, first-serve order. However, some systems provide for different queue discipline alternatives. For example, the system may allow a priority bin to be pulled out of sequence and moved to the head of the waiting line.

Queuing with first-come, first-serve discipline is by far the simplest to program and manage. However, the system program may include provision for deviating from this discipline for priority picks or for other reasons. In addition to the program routines for priority picks and revised sequencing, the system must provide physical space for the bins in the queue. Configurations that provide queuing space were discussed in a previous section.

An important function of the queue is buffering or banking waiting bins so the worker doesn't have to wait for the next bin to arrive at his work position. This may increase the waiting time of the bins, but this is inconsequential to overall system productivity.

15.5.5 Stacker Control

Computer control of the stacker vehicle operates on two levels. The first is the system controller, which issues instructions identifying the the bin and its location for pickup and delivery to the station destinanation. After the bin has been worked, the system controller issues return instructions. The second level is the microprocessor on board the vehicle. This controls such movement details as acceleration-deceleration, accurate positioning for extraction and replacement, and operation of the extraction mechanism.

Communication between the system controller and the stacker vehicle can be in one of two modes. One is communicating with the stacker only when it is at its base position. Once the stacker is in motion, it is out of contact with the system controller until it returns to its base location. The other mode provides continuous communication capability when the stacker is in motion and no matter where the stacker is located. This is more expensive than communicating only at the stacker base position, but does enable the system controller to revise instructions while the stacker is in motion.

15.5.6 Extraction and Insertion

The operation of the extraction mechanism is controlled by a microprocessor on board the stacker. When the stacker is accurately positioned at the bin's rack location, the microprocessor directs the extraction mechanism to pull the bin onto the stacker platform. After the stacker has reached the work station, the extractor pushes the bin off the platform onto the work station. When the pick is completed, the extractor pulls the bin back onto the platform. After the stacker has reached the designated storage location, the extractor pushes it from the stacker platform onto the rack at the proper location. These operations are all directed by the microprocessor on board the stacker.

16

Carousel Storage and Retrieval Systems

16.1 INTRODUCTION

A carousel is a movable loop conveyor for materials handling and storage. It has start, stop, and retrieval capabilities with either manual or automated controls. The conveyor may be at floor level with the product on top of the carousel or elevated with the material hanging below.

The first and still the most common carousel application is storing and handling completed garments in a retail dry-cleaning establishment. The carousel is a moving conveyor from which garments are hung according to a specific identification plan. The clerk stands at the end of the conveyor near the customer service desk and causes the carousel to revolve until the desired garment arrives. It is removed and delivered to the customer.

The carousel eliminates the clerk walking through a series of racks looking for the customer's garment. Putting the completed garments on the rack can be done at a single work station, thus eliminating more walking among racks. It is an efficient space user since little or no clearance is required except at the receiving and discharge points.

The carousel principle is now widely used in materials storage, handling, and retrieval. It requires less capital investment than automated storage and retrieval (AS/RS) and miniload systems, yet produces worthwhile operating savings. Carousels also use less space. A typical one is shown in Figure 16.1.

A common application is storing inventories of small parts and supplies. The storekeeper can call for needed parts, pick them, and

Fig. 16.1 Typical carousel for varied inventory with manual picking. (Courtesy White Storage and Retrieval Systems, Inc.)

call for the next item without walking or moving from the station. A number of stock-keeping units (SKUs) can be stored at one station if the volume of each SKU is small.

Another carousel application is storing work in process. Partially completed assemblies can be placed on the carousel and moved to succeeding positions as needed. Another application combines carousel storage with a test operation. Electrical assemblies are mounted on a carousel and moved into a test station. After the testing is completed, the assembly can be moved on, if accepted, or removed from the system, if rejected.

16.2 ADVANTAGES

Carousels have a number of advantages over other materials handling systems. Their simplicity often leads management to overlook their good points in comparison to highly automated systems such as AS/RS, miniload systems, and automatic guided vehicles.

16.2.1 Space Utilization

Carousels are efficient space users. They do not require extensive aisles since access is needed only at the insertion and retrieval points. Otherwise, the only clearances needed are for fire and maintenance and for the elimination of damaging contacts with material and structures.

Carousels can be fitted into available space. Within engineering constraints such as radius, carousels can be lengthened, shortened, or turned around corners to fit into the space. If more capacity is needed and building height permits, the carousel height can be increased and access gained by a mezzanine or ladder platforms.

16.2.2 Customer Service

Carousels were originally installed to improve customer service as well as to reduce labor costs. This applies whether the customer is external, as in a dry-cleaning retail store, or internal, such as a worker getting tools from a tool crib. The needed items can be retrieved and delivered quickly and efficiently.

Another feature of customer service is the ability of a carousel system to handle orders in any order desired. A rush order can be easily moved to the front position and handled expeditiously.

16.2.3 Installation Cost

Lower installed cost is a major advantage of a carousel system. A common rule-of-thumb is that the capital investment for a carousel is only about half that for a more highly automated system. This is extremely important for an organization with limited capital funds (as are most companies). A corollary to lower investment cost is that usually the delivery and installation period is less for a carousel than for a highly automated system.

16.2.4 Flexibility

Flexibility is an outstanding advantage of a carousel storage and retrieval system. Expanding a carousel system is easier than enlarging an AS/RS, miniload, or other automated system. The installation and investment cost of one more carousel is only about half the cost of rebuilding an AS/RS. If space and volume conditions dictate, existing carousels can be extended. If the size of the expansions warrants, another carousel can be added. Reprogramming the computer controls should present no insurmountable problems, although some time and effort will be needed.

The storage container size, configuration, and arrangement can be altered as inventory requirements change. Bins of different size can be installed at a carousel station. Software must be modified to handle the new arrangement.

16.2.5 Access Points

A carousel system requires at least one access point for extraction and restocking. If only one such point is required, coordinating the two activities may become a problem. One solution is to restock in an off-period, such as during the evening.

A better solution is to have more than one access point. This separates restocking from extracting and substantially reduces the confusion and chance for error. If both activities are going on at the same time, priorities must be set. Usually the extraction and delivery takes precedence. This may cause inefficiency in loading and frustration to the workers doing the restocking. It is better to separate restocking and extraction both in time and in location.

16.2.6 Only One Move per Activity

An operating advantage of the carousel is that only one move per activity is required. The bin called for is moved into position for order picking or other activity. When this is completed, the bin doesn't have to be returned to its previous position or to any exact location. The next bin called for moves into the work station, and the completed bin moves off simultaneously. It doesn't matter where the completed bin goes. It maintains its standing in the carousel queue and can be recalled directly from any position on the carousel track.

16.2.7 Integration of Carousels with Manufacturing Resource and Other Systems

A carousel is not necessarily a stand-alone system. It can be interfaced and integrated with the materials requirements system or with manufacturing resources planning, if desired. This requires feeding issue, receipt, and inventory information to the system controller. This process can be manual or may be highly automated, depending on the system needs.

16.2.8 Better Throughput Through Fewer and Shorter Moves

Originally the carousel operator called for each item individually in the order in which the items appeared on the order sheet. The movement time of the carousel can be reduced by a system controller, which receives the entire order and then calls for each item in an order that minimizes travel distance and consequently time. This increases throughput without interfering with the efficiency of the system.

16.2.9 Reliability

Carousels have higher reliability than do other automated materials handling systems. They are simpler with less to go wrong.

16.3 CONSTRUCTION

In this section, we shall discuss the physical components of the carousel.

16.3.1 Drive

Carousel movement results from power exerted by the drive unit. This is an electric motor which applies power to the carousel conveyor drive chain. Various types of motors are used.

The moving carousel conveyor may be at either the top or the bottom of the carousel stacks. The original configuration was top drive. The items (garments) were hung from the moving conveyor. Since the unit weight was low, there was little problem with structural design and operation.

If the unit weight is high, a top-driven conveyor poses some engineering problems in strength and deflection. One way to solve these problems is to mount the conveyor in the floor where the conveyor elements can be supported by the floor. This permits the carousel to transport heavier loads.

16.3.2 Power Source

Power to run the carousel is supplied by the plant's electric system. Motors at both ends of the carousel are required for heavy loads and long conveyors. Alternating current (AC) is usually cheaper and doesn't require any transform from the line current. AC drive is good for continuous operation, but does require a clutch to protect the drive mechanism and to avoid jerky starts if the operation is intermittent.

Direct-current (DC) drives are more expensive than AC drives but are preferable under the following conditions:

1. Smooth starts and stops are needed.
2. Stopping accuracy tolerances are strict.
3. Speeds and loads are variable.
4. The system uses robots for transferring parts.

DC drives are more flexible in programming speeds and changes in speeds.

16.3.3 Controls

The carousel storage and retrieval system requires a mechanism for controlling and directing the operation of the system. Three major types are available — manual, microprocessor, and computer. The choice is determined by the system's needs and the capital available.

The simplest and cheapest is a manual control. Among the types available are:

1. Foot control. A foot-operated pedal activates the carousel in the desired direction. Normally, this requires the operator to release the pedal to stop the carousel when it arrives at his work station. Close stopping tolerance is not feasible with manual control. An approved guard for the pedal must be provided.
2. Hand control. The worker operates the carousel by a hand switch. This is simple, but requires the operator to wait and observe the carousel movement to stop the desired bin at the work station.
3. Keyboard control. The worker inserts the call for the desired carousel into a keyboard terminal. The system delivers the bin to the work station by the shortest route. This control mode also provides other capabilities, such as cathode-ray-tube display and emergency stops.

Modern carousel installations usually have some form of microprocessor control. It may be an interface box, a microprocessor that communicates through a network with other carousels, the host computer, the system controller, on-site terminals, and other components of the system. It provides accurate tracking and stopping plus the ability to interface with other devices and to perform diagnostic routines.

Another type of microprocessor has a memory. This provides capability to monitor more than one carousel, to queue requests according to predetermined queue discipline, and to handle a large number of requests at one time. Bin locations are stored in memory and available for intelligent choice.

A dedicated minicomputer is recommended for most carousel installations. It provides real-time operational control, an essential feature of efficient operation. The system controller can interface effectively with any microprocessor or programmable controller in the system. Many standard software packages are available for computer operation of a carousel system.

16.3.4 Conveyor

Most present carousel systems for picking orders use manual loading into the carousel bins for belt or roller conveyors. The same conveyors are used for moving completed orders out of the system. Another possibility is an overhead-powered conveyor. Some systems use transporters with tote boxes. In most of these systems, an operator can handle both the carousel for order picking and the transporter for incoming product and outgoing orders.

16.3.5 Inventory Containers (Bins)

Bins are made of a variety of materials. Wire and plastic are popular because they are light. Standard sizes should be used when possible to avoid the expense of special design and construction in the carousel system. Bins that carry heavy loads, 75 lb or more, are made of steel, strong plastic, or are reinforced if made of wire. Bins are placed in position on the shelves usually without any catch attachment (see Figure 16.2).

Shelves are provided to hold the bins. Usually they are tilted slightly backward to hold the bins and parts in place as the carousel moves. Each shelf can have a back and two sides to hold the bin firmly in place as the carousel revolves. Wire racks for holding bins and shelves are illustrated in Figure 16.3

Special bins and shelves can be used. Unlike some other features of automated materials handling systems, these are not excessively expensive in comparison to their value to the system if they will fit standard carousel dimensions. One example is a special fixture to hold electrical assemblies in a test facility in which a carousel moves the assemblies into and out of the test position.

16.3.6 Automatic Load-Unload Divices

Automatic insertion and removal devices can be used with a carousel storage system. These require a solid bin made of metal or plastic. A wire container is not suitable. This feature requires computer

Fig. 16.2 Carousel with compartmented bins. (Courtesy White Storage and Retrieval Systems, Inc.)

Fig. 16.3 Carousel with wire racks. (Courtesy White Storage and Retrieval Systems, Inc.)

controls and accurate stopping tolerances. The movements and stops of the carousel must be coordinated with the insertion mechanism.

16.4 APPLICATIONS OF CAROUSEL STORAGE AND RETRIEVAL

In this section, we shall report on and discuss several representative types of carousel application. Many more are possible. New ones are developed as users gain experience. The carousel is versatile, flexible, efficient, and inexpensive to build and operate. These characteristics have led to its popularity.

16.4.1 Inventories

The most common type of carousel installation handles inventory. This is an efficient space user and provides easy access for order picking, kitting, or other functions.

Finished-Goods Inventory

Carousel systems are well suited to finished-goods inventory applications. Finished product is brought to the carousel in unit loads or in lots. Orders of almost any size and complexity can be picked from the carousel. This is good for distribution centers that ship in small quantities commonly less than unit loads. Other systems are better for distribution centers shipping in unit loads, case lots, or truckloads.

Work-in-Process Inventory

Some carousel installations handle work-in-process inventory. This is useful for partially finished product which must be temporarily stored pending a call for the next operation or for completion. This type of system usually has tracking capabilities, so the location of individual items is stored in the data bank and the desired item can be called when needed. (See Figure 16.4).

Raw-Materials Inventory

Carousels can be used to store incoming raw materials if the size and volume of the items permit. An example is a raw-materials inventory consisting of small parts and small assembled components. An automated carousel system can store these items efficiently and can keep track of their location and status for ready retrieval. A carousel system for small parts can be combined with another type of raw-materials storage for items too large or too numerous for carousels.

Fig. 16.4 Carousel for computer terminal work-in-process. (Courtesy White Storage and Retrieval Systems, Inc.)

A few installations receive and store raw materials as received from vendors. This is practical when there are large numbers of raw-materials SKUs that must be tracked quickly and accurately. There are other systems more efficient for handling incoming material that has only a few SKUs.

Some systems combine raw-materials and work-in-process carousel storage. This is appropriate when similar items are in both inventory catagories. A manufacturer may purchase some components from outside vendors at the same time he is making some similar items in-house. Even though the accounting classification may be different, the physical characteristics and system requirements are similar enough that both types can be stored in the same system.

Supplies and Maintenance Spares

Carousels are often ideal for storing office and shop supplies and maintenance spare parts. The volume of these items is usually small to moderate, so space requirements are modest. They are usually small enough to fit into carousel bins. The number of SKUs ranges from moderate to large. An automated carousel system can keep good track of items in the inventory. This is important for equipment spare parts since a delay in locating a vital spare part can be costly in lost production.

Tools

A difficult storage problem in many factories is small tools needed for production. A carousel is good for this purpose since it can be operated effectively by one tool crib attendant. It also can keep track of the status and location of a large number of small items. However, the tool inventory must be purged periodically of items no longer needed. Otherwise, valuable space will be occupied by tools for which there is little or no demand. Another problem may be weight; some tooling, such as punch and die sets, may be too heavy for economical carousel storage.

16.4.2 Electrical Testing ("Burn-in")

One interesting application of carousels is testing electrical and electronic assemblies. The assembly is placed in a carousel bin. As the carousel revolves, the assembly reaches the test station, where it is automatically connected to the test equipment. After the test is performed, the results are reported, and the tested assembly moves out of the test station. The system must have provision for removing unsatisfactory assemblies. Although it is possible to do this automatically, the usual practice is for the system to send a signal to an operator who manually removes the defective assembly.

16.4.3 Buffer Storage

A carousel can be used to advantage for buffer storage in an assembly operation. Partially completed assemblies are inserted into carousel. They keep in motion until removed by an operator needing work for the next operation. This provides a buffer bank of partially completed work handy to the assemblers so there is no operator delay getting work for the next operation. It is also relatively economical in floor space over other banking methods.

16.4.4 Flexible Manufacturing System

The flexible manufacturing system (FMS) is a new, completely reorganized manufacturing methodology. It combines several elements previously not integrated into a single system. The elements are computer control, automated materials handling, numerically controlled machine tools, and robots. FMS are further described in Chapter 18.

A carousel storage and retrieval system works very well in an FMS. It has the advantages of limited space requirement, automatic control, automated extraction and insertion of product, and the capability of handling raw materials, work in process, and tooling in the same system. Economical space use is important since many FMS are in older manufacturing space, which may be limited. The carousel is usually integrated with other types of automated materials handling systems. For example, materials coming into or going out of the system may move on conveyors while the carousel is used for temporary storage of product and tooling.

16.4.5 Progressive Assembly

A carousel can be used between successive stages of a progressive assembly operation. The partially completed work is inserted into a carousel bin and remains there until called for by the operator. This is useful when a number of different models are produced on the same assembly line since there is no difficulty in calling the proper work-in-process assembly as it is needed.

16.4.6 Sorting and Consolidating Returned Goods

Some industries have a large number of returned items to process. An example is a company that gets a steady stream of items returned for repair or modification. Another is a distribution center that gets many returns from customers for various reasons. In each case, the returns must be processed and directed to the correct destination.

A carousel can be used in the receiving operation. The receiving clerk inspects, identifies the destination, prepares the necessary receiving ticket in either hard copy or computer terminal entry, and inserts the item into an empty carousel bin. He also enters the bin location on the receiving document or entry. The system controller takes over and moves the returned item to its destination either back into inventory or to a repair station.

16.5 DESIGN PRINCIPLES

In this section we shall discuss some important design features of a carousel installation. It's not possible to discuss all design characteristics; we include the most important ones that are involved in carousel system design.

16.5.1 Length

Standard carousel layouts range up to 100 ft in length. The shorter the length, the shorter the replenishment and retrieval times. On the other hand, the shorter carousel requires almost as much investment in power drives, controls, and other features as does a long one. A longer carousel may be limited in accessibility through limitation on the number of access points.

The area available often limits the length of the carousel. Many companies chose a carousel over other materials handling systems because of space limitations combined with inability or unwillingness to expand the building or build a new facility.

16.5.2 Height

The height of the carousel is limited by the following:

1. The height of the available space.
2. The weight of the carousel and the inventory carried. Since the carousel moves, engineering problems of statics and dynamics limit the height. The stationary rack system, such as used in an AS/RS, can go to much greater heights.
3. Access to the carousel containers. Most carousels use manual picking and insertion. The worker must be able to reach the bins efficiently. A mezzanine may facilitate effective use of a multitiered carousel, but this has its limits.

A carousel installation may consists of separate carousels one above the other. They may have a common drive or separate ones with separate controls for each tier.

One solution to the problem of height is a movable or stationary ladder for an operator. Although this is operator inefficient compared to having a lower carousel with bins accessible to an operator standing on the floor, it may be the best combination of factors. The productivity of this configuration will be enhanced if the system controller can assign slow-moving items to higher locations. An example is shown in Figure 16.5.

Fig. 16.5 Multiaisle carousel with operating terminals and movable ladder. (Courtesy White Storage and Retrieval Systems, Inc.)

16.5.3 Horizontal or Vertical

Most carousels operate in a horizontal plane. This is best from engineering and operating points of view. A horizontal carousel requires less investment for a given capacity.

Vertical carousels are useful in special situations. The most common application is storage and retrieval of documents such as medical records and x rays. Vertical carousels are superior in security and in locations where floor space restrictions are important. A vertical carousel may be subject to imbalance, a condition that seldom occurs with a horizontal one.

16.5.4 Drive Location

The drive moves the carousel. A major decision is whether to have the drive at the top or bottom of the carousel. Originally most carousels were top driven. Bottom location has become popular for two reasons: A bottom drive can handle heavier weights more reliably. Also, there is little danger of contamination from dripping oil or other substances.

The drive motor size will depend on several factors. The most important is the weight of the installation. Another important consideration is the ability of the drive to accelerate and decelerate rapidly to minimize travel time.

16.5.5 Mezzanine

The high cost of building floor space makes building up rather than out economical. Although the practical height of a carousel is limited to 15–20 ft, this is still too high for an operator to pick manually from all bins. The solution is a mezzanine which permits order picking and other operations on two levels. The mezzanine may be built only at the pick stations for economy or space reasons, although it is better to have a mezzanine all the way around the carousel if space permits. Otherwise, there may be difficulties in emergencies and in maintenance and servicing access.

16.5.6 Aisles

A major advantage of carousels is the reduction in aisle space required. Access points can be located at the ends of the carousel. This permits the carousel to be located within inches of the next one if no maintenance or emergency access is required. An emergency or service aisle seldom needs to be over 2 ft wide. This is far less than a system that requires product to be removed and inserted from into racks from the aisle.

16.5.7 Continuous or Intermittent Movement

Most carousels operate intermittently or only when activated to load or unload an item. When activated, the carousel revolves until the desired bin arrives at the operator's work station. Most control systems can move the carousel by the shortest route from start to destination. Activation can be manual or by microprocessor or computer.

Some installations use continuous carousel operation. An example is an assembly operation in which an operator removes an item from a slowly moving carousel, works on it, and replaces it in an empty space. Moving carousels can be used for testing or burn-in-operations.

16.6 COMPUTER-CONTROLLED CAROUSELS

Computer controls for carousels cost more than manual ones but have several advantages which more than offset the added cost. Each installation must be individually examined to determine the most efficient and economial way to meet the company's needs and constraints.

16.6.1 Random Storage

In random storage, inventory locations are not assigned to specific parts or parts families. Instead, the computer selects an open slot and directs the inventory to be stored in that spot. At the same time, the computer keeps track of where each inventory item is stored. When the item is called for, the computer determines its location from the updated database and directs its removal from that location.

Random storage increases storage density. There is no need to hold specific locations for a specific item. This practice results in empty locations. In random storage, the computer can make full use of vacant spaces. Since it knows where each item is stored, it can recall a desired item without trouble and delay. It is also possible to program the location assignments by product size and quantity if a standard unit loads can not be used for all products.

16.6.2 Sequenced Picking

The computer control can sort requests by location and arrange them in an order to minimize travel distance and time. This cuts the waiting time between picks. If the operator can pick from more than one carousel at his work station, he can pick from one while another bin is moving into the position.

16.6.3 Multiple Picks

When several orders are to be picked at the same time, computer control helps the operator to pick items from one bin for several orders. The computer accomplishes this by arranging the orders to be picked so the operator can identify and make multiple picks. This reduces travel and waiting time. If item demand follows Pareto's law, a large number of line picks will come from a small number of bins. This further improves the efficiency of the multiple-pick system.

16.6.4 Faster Replenishment

Random storage, possible only with computer, eliminates the need to search an assigned location that is empty. The random-storage routine chooses the best available empty location.

16.6.5 Reduced Inventory

With a computer-controlled carousel system, inventory can be lowered by using the computer-generated information on inventory status. The system furnishes real-time perpetual inventory data. This enables the company to avoid overstocks and stockouts while reducing the inventory investment.

16.6.6 Better Physical Control

The carousel limits access to inventory to the replenishment and retrieval points. This provides better physical control of the inventory. Also, the bins are visible so that unusual inventory situations can be noted by the supervisor.

17

Automatic Guided Vehicle Systems (AGVS)

17.1 INTRODUCTION

17.1.1 Definition

Automatic guided vehicle systems (AGVSs) feature battery-powered driverless vehicles with programming capabilities for destination, path selection, and positioning. The vehicles are similar in construction to driver-operated ones with modifications for automatic operation. Each vehicle also has provision for features necessitated by the absence of a human operator.

The modern AGVS depends on the modern digital computer. An on-board microprocessor guides the vehicle along the prescribed path and makes corrections if the vehicle strays from the path. Each AGV is in communication with a system controller which receives instructions directly or from the host computer, correlates the instructions with status information in its database, and issues appropriate move commands to the vehicle. Without the modern computer, the system would be too slow, expensive, and cumbersome to operate efficiently.

17.2 AGVS COMPONENTS

The AGVS has four main components:

1. The vehicle, which moves the material within the system without a human operator.
2. The guide path along which the vehicle moves.

3. The controls, which direct and monitor system operations. This includes feedback on moves, inventory, and vehicle status.
4. Interfaces with other computers and systems such as the organization's mainframe host computer, an automated storage and retrieval system (AS/RS), or a flexible manufacturing system (FMS).

17.3 VEHICLES

Several different types of vehicles are available to meet service requirements. The basic vehicle is a modification of existing nonautomated equipment. Most are mechanically rugged and reliable.

17.3.1 Towing

An AGV can be used to pull a train of industrial trailers. An example of an AVGS towing vehicle (tugger) is illustrated in Figure 17.1. This is economically and energywise more efficient than carrying each unit load on a separate AGV. A wide variety of trailers can be used, such as hand pallet trucks, custom trailers, bin trailers, or flat beds.

There are several different types of equipment for loading. AGV-pulled trains, fork-life trucks, hand pallet trucks, cranes, automatic transfer equipment, manual labor, and shuttle transfer are in use. Automatic loading and unloading can be programmed.

Industrial trains, whether operator-driven or AGV-pulled, are seldom efficient for short distances. They are better when large volumes must be moved over distances of 1000 ft or more.

17.3.2 Unit Load Transporters

An AGV unit load transporter is one which carries an individual load on board the vehicle. They are quite versatile and can be used effectively in a wide variety of situations. Automatic load and unload capability can be provided, as can bidirectional movement. Figure 17.2 shows a typical unit load transporter.

An AGVS unit load transporter can be furnished with lift and lower tables to pick up and deliver loads from different heights of conveyors and load stations. This is important for an AGVS that is part of a larger production or distribution system where the AGV must interface with other materials handling system and equipment.

A unit load transporter can be built with a conveyor platform to interface with powered or unpowered conveyors. Push-pull stations can be incorporated. Loads can be inserted and removed by fork-lift truck or other device.

Fig. 17.1 An AGVS towing vehicle. (Courtesy Barrett Electronics Corp.)

Fig. 17.2 Unit load transporter. (Courtesy Barrett Electronics Corp.)

17.3.3 Pallet Trucks

Originally, AGVSs used existing pallet trucks with AGV capability added. They are still widely used. This vehicle can be manually operated when loading or unloading. It is useful for picking up and dropping loads from floor level. No special device is necessary to lift, load, or unload an AGV pallet truck. An automated guided pallet vehicle is pictured in Figure 17.3. One with pallet-lifting capability is shown in Figure 17.4.

17.3.4 Assembly Lines

An AGVS can be used as the basic move mechanism of an assembly line, such as one for automobiles. The AGV has a special fixture on which the vehicle base is mounted. As the AGV moves from one station to the next, succeeding assembly operations are performed.

Fig. 17.3 An automatic guided pallet vehicle. (Courtesy Barrett Electronics Corp.)

Fig. 17.4 An automatic guided pallet vehicle with lifting capability. (Courtesy Barrett Electronics Corp.) Note: Unicar and Guide-O-Matic™ are registered trademarks of Barrett Electronics Corp.

A major advantage of this system is the lower expense and ease of installation over "hard" assembly lines. The line can be changed by moving the guide path if necessary and by reprogramming. Furthermore, variable speeds and dwell intervals can be programmed into the system.

17.3.5 Light Load Transporters

These are AGVs designed to move lighter loads than the usual production and distribution ones moved by previously described AGVs. Their primary use is for moving parts, supplies, and mail in light manufacturing. Usually they are loaded and unloaded manually.

17.4 GUIDANCE

An essential component of an AGVS is the guide path and guidance system. This is the mechanism that keeps the vehicle on the designated path. There are no chains, cables, conveyors, or other fixed devices. The two types, optical and magnetic, operate by signals which keep the vehicle on track and return it to the path if it deviates.

A key advantage of AGVS is the low cost of changing the guide path in comparison to the cost of modifying fixed-path equipment such as conveyors, chains, or tow lines. Even cutting into the floor for the magnetic system wire is easy and inexpensive. Furthermore, the guide path does not obstruct other traffic.

17.4.1 Optical Guidance Systems

Optical guidance systems depend on reflective tape or painted stripes on the floor. The AGV directs a light beam onto the guide path, which reflects the beam back to the AGV. If the light amplitude falls off, the on-board control system sends out commands which return the AGV to the path.

Optical guide paths are useful in office or clean environments, for a temporary guide path, or for guide paths that must be changed frequently. They require a high reflectivity contrast between the guide path and the unmarked floor. Consequently, they must be kept clean and replaced frequently because of wear. Optical guide paths are not suitable for harsh industrial environments.

17.4.2 Magnetic Guide Paths

In this system, a wire is placed in a slot cut into the floor. The slot is filled to preserve the floor level. The wire emits a low-current, low-frequency AC signal that produces a magnetic field around the wire. The AGV carries sensors that detect this magnetic field. If the vehicle deviates from the guide path, the AGV's on-board microprocessor directs action to return the vehicle to the guide path.

Magnetic guidance is almost completely maintenance-free. It can be used in harsh industrial environments with very high uptime.

17.5 AGVS CONTROL SYSTEMS

The essential difference between an AGVS and a operator-driven system is the sophistication of the controls required in the absence of the human operator. The human being is a complex, flexible organism with high levels of sensory and decision ability. Many functions an operator performs almost with attention or thought must be programmed into the AGVS.

There are three levels of AGVS control:

1. Computer control
2. Remote dispatch
3. Manual control

17.6 COMPUTER CONTROL

Computer control of an AGVS is the most efficient mode, but it is also the most expensive and complex. In this mode, all transactions and vehicle movements are directed and monitored by the system controller. A diagram of a complex system is shown in Figure 17.5.

17.6.1 Computer Levels

As discussed in Chapter 12 "Computer Developments," there are three levels in the system that operates a AVGS:

1. Microprocessor
2. System controller
3. Host computer

Microprocessor

A microprocessor is a small computer programmed to perform a limited number of functions. Its small size allows it to be placed at the point of use. For example, a microprocessor can be put on the vehicle to control its adherence to the guidepath and to correct any deviation from that path. It can also operate the automated load/unload mechanisms. Originally, microprocessors had fixed programming and could not be reprogrammed. Consequently, if reprogramming were necessary, a new microprocessor, or at least new circuitry, had to be installed. The latest development is the programmable controller, a microprocessor that can be easily reprogrammed rather than replaced.

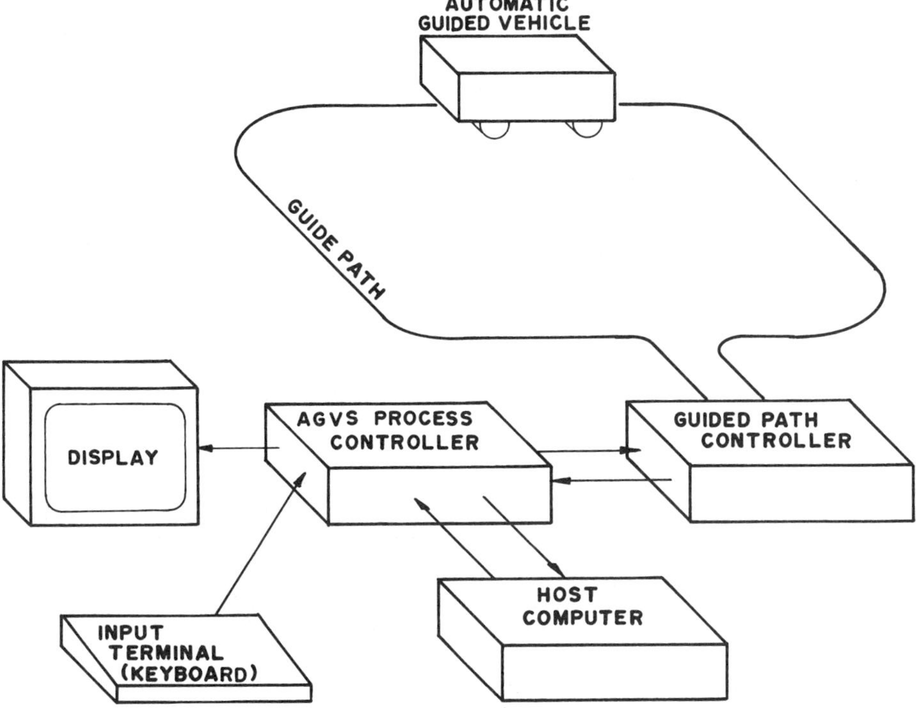

Fig. 17.5 Complex automatic guided vehicle system (AGVS).

System Controller

The key to AGVS operations is the system controller. This is a mini-computer that handles the instructions to the vehicles and collects information on system moves, vehicle location, and inventory status. It deals directly with the vehicle microprocessors. It also sends certain data to the host computer. Only pertinent data are forwarded to the host. The rest remains within the system controller and periodically is purged when obsolete. Typical system controllers are Digital Equipment Corporation's VAX ® and PDP ® series computers.

Host Computer

The host computer is the organization's central computer mainframe. This computer processes accounting, financial, and other data important to company management and to the accounting department. These data come either from the system controllers or from direct entry. The host computer can also issue overall production plans to the factory. It normally does not concern itself with operating details of the individual systems such as the AGVS.

17.6.2 Dispatch Instruction

When a transaction instruction is entered into or received by the system controller, the controller selects the best (usually the closest) vehicle to make the move. It issues instructions to the vehicle, which then carries out the command. In some systems, the destination address is determined by the computer. In others, the destination is entered manually. In the latter case, the system edits the manual entry to ensure validity and to eliminate human error.

17.6.3 Automatic Load/Unload

Almost every computer-controlled system includes automatic load/unload capability. Otherwise, the manual load/unload operation would require too much labor and expense to make the system economically viable. The automatic load/unload feature also increases efficiency by reducing idle vehicle time.

17.6.4 Interfaces with Other Systems

The computer control facilitates interfacing the AGVS materials handling system with other systems in the factory. These include AS/RS, computer numerically controlled machines, FMS, process control equipment, and shop floor control systems. This interface may be through the host computer, although it is more efficient to use a distributed data-processing network. In a distributed network, the system control computers communicate with each other directly rather than through an intermediate or host computer.

As the number of systems interfaced rises, the damage potential of failure of any part of the system rises. If the failure of the AGVS will shut down production for a period, the lost production cost could be substantial. This loss possibility has motivated systems designers to improve the reliability through better design, higher quality standards, and redundant circuitry.

17.7 REMOTE DISPATCH

In this mode, the human operator issues instructions through a remote station rather than dealing directly with the vehicle. The

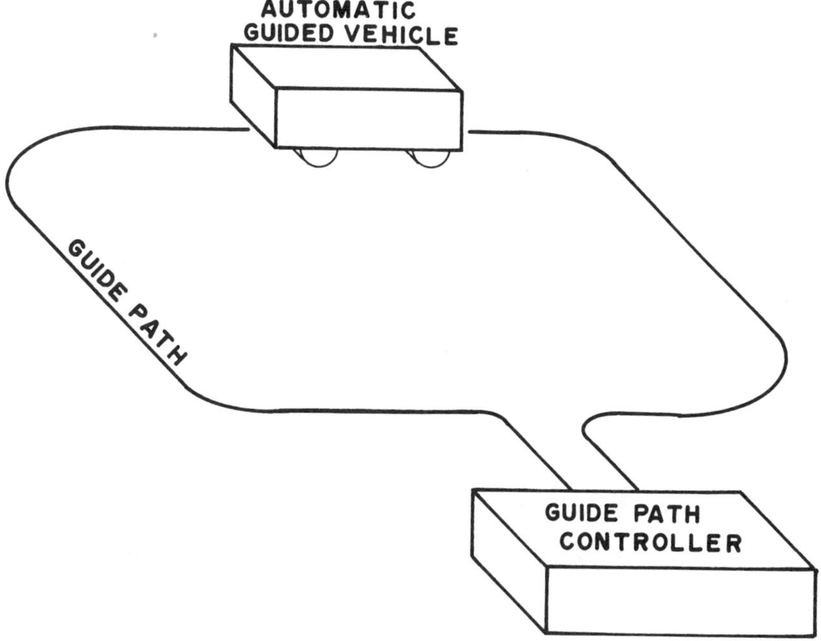

Fig. 17.6 Simple automatic guided vehicle system (AGVS).

system sends destination instructions directly to the vehicle. Figure 17.6 sketches a simple system diagram.

Remote dispatching does not provide for computer feedback on completed moves or vehicle status. It is difficult for the operator to improve system efficiency by picking the nearest or most efficient vehicle for a move. Most of these systems are designed to return the vehicle to a central area to await the next move.

Remote dispatching is more efficient than manual since the operator can enter instructions without waiting for a ready vehicle. Instructions are queued until a vehicle is free and available for the move.

Most remote-dispatching systems use automatic load/unload. Without this feature, there is little advantage over manual control. The station can queue up to 15–25 waiting instructions.

This system starts with a station which receives a load into the system. The empty vehicle picks up the load automatically and proceeds to the designated destination, where it is automatically unloaded. The vehicle is then free for the next move. Usually, it returns to the beginning station to pick up the next load.

17.8 MANUAL CONTROL

In a few automated systems, the operator controls and directs the vehicle dispatching. The operator loads the vehicle and enters a destination into the vehicle's on-board control panel. The vehicle routes itself to the designated destination on the guide path, stops, and awaits further action. The operator must perform or direct the unloading.

Manual control is the least expensive and by far the simplest. However, its efficiency is highly dependent on the skill and performance of the operator. If the worker causes errors or does not keep the system moving efficiently, there will be little advantage over ordinary manual systems. Also, manual systems have no automatic feedback capabilities. The system cannot keep track of the vehicle or inventory status.

17.9 AGVS DESIGN FEATURES

Several design features of an AGVS are common to all materials handling systems; others are special or even unique. In this section we shall discuss some design features.

17.9.1 Stopping Accuracy

This varies considerably with the nature and requirements of the system. It ranges from ±0.001 in. for machine tool interfaces to ±3 in. for a manual system. Towing vehicles and light-load transporters have a tolerance about ±1 in. or more. Unit load transporters are used for applications requiring greater accuracy. If accuracy is necessary, it must be built into the system either by mechanisms or by computer control and feedback routines.

17.9.2 Automatic Load/Unload

An automatic device to load and unload the AGV is essential for top efficiency. If a manual worker is necessary for the load/unload function, there is little point to automating the rest of the AGVS.

A powered roller conveyor on the deck of the vehicle is a common mechanism for the automatic load/unload operation. The vehicle is positioned in front of the station with the system controllers verifying the location and ensuring that the vehicle is correctly aligned with the station. The powered roller conveyor on the vehicle is then activated and the load is moved.

Another mechanism is a lift-and-lower device on the vehicle deck. The load is moved in the raised position. It enters a stand, aligns itself, and lowers the deck to deposit the load on the stand. This method can be used to transfer loads to other conveyors in an interfacing system. Other devices include a shuttle mechanism and a push-pull station to transfer the load to or from the vehicle.

17.9.3 Facilities

Several facilities features must be considered in designing an AGVS. Among them are:

1. Grades. The lower the grade, the better. However, AGVS vehicles can operate with grades of 5% and under special circumstances and go to a 10% grade.
2. Automatic doors. System efficiency is incompatible with manual or remote-dispatch door opening and closing. Automatic controls that open the door as the vehicle approaches and close it after the vehicle has passed are simple, cost-effective, and safe. These controls interface with the vehicles and open and close the door in response to signals from the vehicle.
3. Elevators. AGVSs can operate with elevators between floors. Although this is inefficient, as are almost all elevator installations, it can be implemented in a multistory building which offers no other alternative.
4. Environment. The environment must be free from conditions that impede the AGVS. Among these are excessive noise and electrical interference.

17.9.4 Safety

The AGVS must incorporate a number of safety features to avoid injury to workers and damage to inventory, equipment, and buildings. Some safety features are:

1. Emergency contact bumpers. The vehicle is programmed to stop immediately on contact. Bidirectional vehicles will have a bumper on each end. These devices with the accompanying controls and brakes are sensitive enough to stop the vehicle without injury to a person or damage to property.
2. Automatic signals to warn workers of the vehicle's approach. Flashing light and audible tones are two common devices.
3. Emergency buttons enable workers to stop the vehicle in a very short distance.
4. An automatic stop feature is similar to a dead-man's switch. This stops the vehicle immediately upon loss of power or other predesignated occurrences.

5. Object detectors which stop the vehicle when it approaches too closely to a person or object. These operate with ultrasonic or infrared radiations.

17.10 APPLICATIONS

In this section, we shall discuss examples and applications of AGVSs to materials handling problems. Some were developed in Europe and Japan before they were introduced in the United States. New applications are being developed as technology improves and as experience is gained.

17.10.1 Raw-Materials Storage

Some of the earliest applications of AGVs were in moving material over long distances within the facility. For example, raw materials were stored at a warehouse over 1000 ft from the production floor. There may be sound reasons for separating the raw material inventory from production. Manufacturing space may be limited and difficult to expand to provide inventory room. Raw materials could be stored in structures that were less expensive to build and operate. For example, storage temperatures could be much lower than those required for production. AGVs are suited to this situation since during much of the trip, the AGV can be put on automatic with a substantial saving in labor cost of industrial train drivers. The AGV was a tugger tow vehicle, one of the earliest and simplest types made. The carts were unchanged. Loading and unloading was usually done by a fork truck. However, the fork trucks could be kept busy at each end of the route instead of spending considerable time going from one end to the other.

17.10.2 Finished-Goods Storage

In many manufacturing plants, the finished-goods storage and distribution facility is separated from the production area. This separation may have sound reasons, such as clean environment, protection against damage and pilferage, or manufacturing space too valuable or limited. Wide separation may be the best use of existing buildings.

Again, the AGV was usually a tugger towing vehicle that moved large quantities of material from manufacturing to finished-goods storage. The original installations used fork trucks at each end. At the inventory end, the fork-truck operator had to remove the load from the cart and take it to the designated storage location. This system required permanently assigned storage locations for each item or expensive manual record keeping of quantity and location of each item. However, savings were substantial over the previous method of having the fork-truck operator make the entire trip. Also, installation was easy; it could be done over a week-end or short-time shutdown. Manual controls were used. They were simple and adequate to the task.

17.10.3 Assembly

Recent technological advances in control systems and control hardware have led to the use of AGVs in assembly operations. This approach has many advantages:

1. The AGV can be programmed to move at variable speeds or on operator command. This introduces flexibility into the assembly sequence and tends to alleviate the difficulties and unpleasantness that come with an assembly line moving at a fixed speed.
2. The installation cost is far less than that of a "hard" assembly line.
3. The programming can be revised often without disturbing the physical configuration of the equipment. In contrast, revising a "hard" assembly line is a major redesign-and-installation project.

17.10.4 Flexible Manufacturing Systems

A major element of the FMS center is the materials handling function. Raw material or partially finished product must be moved into the center and placed in the machine. It must be moved from machine to machine within the center; this is often done with robots. When the part is completed, the part must be moved out of the center to storage or other destination.

AGV are now being used for moving the product into and out of the FMS center. They have the advantages of

1. Flexibility in routing. The control program can provide as many different routing patterns as needed. Since routings are in the control computer, no hardware or building change is needed for different products. New routings can be added and obsolete ones deleted easily.
2. Accurate positioning. Modern control technologies provide accurate stopping within the tolerances required by the machining processes.
3. Lift and lower capability. The AGV transporter can raise the material to the level of the machine tool and can remove the material from any machine tool height within the AGV capacity.

17.10.5 Manufacturing Operations

One manufacturer uses AGVs for transporting material through its plant, making a key assembly. AGVs pick up the incoming material from an (AS/RS). The AGV transporter moves along the guide path until it reaches the machining station, where the material is automatically unloaded and moved to the machine tool. When the work is completed, it is moved back the same way to the out station. The transporter moves the material to each successive station, which can be another machining operation, an assembly, or storage. Computer controls are used to generate a production schedule and to issue commands for materials movement according to the schedule. Separate control units oversee the traffic flow.

17.11 ADVANTAGES

17.11.1 Increased Reliability

AGVS have several reliability advantages over other automated materials handling systems:

1. If an AGV breaks down, it can be pulled from service for repair. Another vehicle can replace the disabled vehicle with negligible interruption in service. In contrast, a system that has a track vehicle dedicated to one aisle may have the whole aisle out of action until the disabled vehicle is repaired.
2. An AGVS is less subject to environmental interference. A magnetic system with an in-floor wire is impervious to environmental contamination such as oil spills, dirt, and noise., Small objects on the floor will seldom deflect the vehicle enough to interrupt operations. A large object will activate the safety bumper and stop the vehicle.

17.11.2 Lower Investment

Since the only track is on a wire in the floor or a reflective stripe on the floor, investment cost is less than that of many other automated systems. The vehicles are produced in large enough numbers to justify economics of scale and consequently lower prices. The big expense may be in the control system and its software.

17.11.3 Flexibility

A major advantage of AGVSs is their flexibility in comparison to other automated systems. Flexibility can occur in several ways:

1. Better use of existing space. It is easier to fit an AGVS into existing space than it is to fit a fixed conveyor or assembly line into the same space. The vehicle-turning radius is small, and there is no hardware or construction which may be difficult to fit in.
2. Changes in the AGVS are relatively simple. The wire guide path in the floor can be relocated by cutting a groove into the floor, inserting the wire, and filling the groove with an epoxy. This can be done over a week-end or other period when the plant is not operating. The control program can be revised without interfering with operations.
3. Often a change requires only reprogramming with no relocating of the guide path or modification of the hardware.

17.11.4 Operating Savings

Usually, an AGVS is cheapter to operate. Less labor is needed than with a manual system. Maintenance is easier and less expensive. Downtime seldom occurs. Power is usually a neglible expense factor. Battery recharging and replacement are the major power expenses.

17.11.5 Inventory Control

Accurate inventory control is a major advantage of AGVS. One speaker at the 1984 National Material Handling Forum reported that lost inventory items were down to one or two per hundred thousand transactions with an AGVS since installation. This is a phenomenal reduction from the number of lost items with a manual inventory system.

The AGVS reporting tracks inventory accurately in real time. This eliminates the motivation for production personnel to order far ahead or extra "just in case." Inventory reductions of as much as 50% have been reported from an AGVS.

17.11.6 No Floor Obstructions

Since the guide path is embedded in or painted on the floor, there is no obstruction to the free movement of personnel and other vehicles over the guidepath. This allows narrower aisles and multiple use by fork-lift trucks and other variable-path vehicles. It also facilitates moving machines and other work that would otherwise require temporary dismantling of a conveyor or fixed-path system.

17.11.7 Interfacing with Other Systems

An AGV can be designed to interface with another automated system, for example, an FMS. The AGV can move material into the FMS and take it out when it is completed. The AGV can deliver the work to the machining station or to an FMS materials handling system such as a robot. Another example is an AGV that delivers unit loads of product from a distant warehouse to an AS/RS or miniload system for order picking and distribution.

18

Flexible Manufacturing Systems

18.1 INTRODUCTION

18.1.1 Definition and Description

The flexible manufacturing system (FMS) was developed to meet the need for automating small-batch production and assembly operations in American industry.

Hard or fixed automation is designed for high-volume mass production of identical or similar products. It is not suitable for small batches of dissimilar products. On the other hand, traditional job shop methods of making small lots of parts are far too expensive, slow, and cumbersome. Many products don't have and never will have the volume to justify hard automation.

The elements of an FMS are:

1. A numerically or computer-controlled machining, processing, or assembly center. Usually, this is a single machine tool, although several may be linked together to form a single center.
2. A mechanism, usually a robot, to move the parts into, around, and out of the center.
3. A computer-controlled materials handling system to move the materials to and away from the center. Automatic guided vehicles (AGVs) are popular for this function.
4. A system controller to issue appropriate commands to coordinate the activities of the FMS and the system components.

In Figure 18.1, we see a diagram of a flexible machining center with one machine tool. The raw material is automatically transferred from the incoming transfer vehicle to the machine tool. After processing is completed, the work is automatically moved to the outgoing transfer vehicle. By having separate tracks for incoming and outgoing product, control and programming are simplified. Figure 18.2 depicts a flexible machining center with several machine tools and only one track. The product is transferred directly from the transfer vehicle to the first machine tool. A robot moves it from machine to machine until it is completed and then transferred automatically to the outgoing cart. Although one track is less efficient than two, it may be satisfactory if the volume is low and imperative if available space is limited.

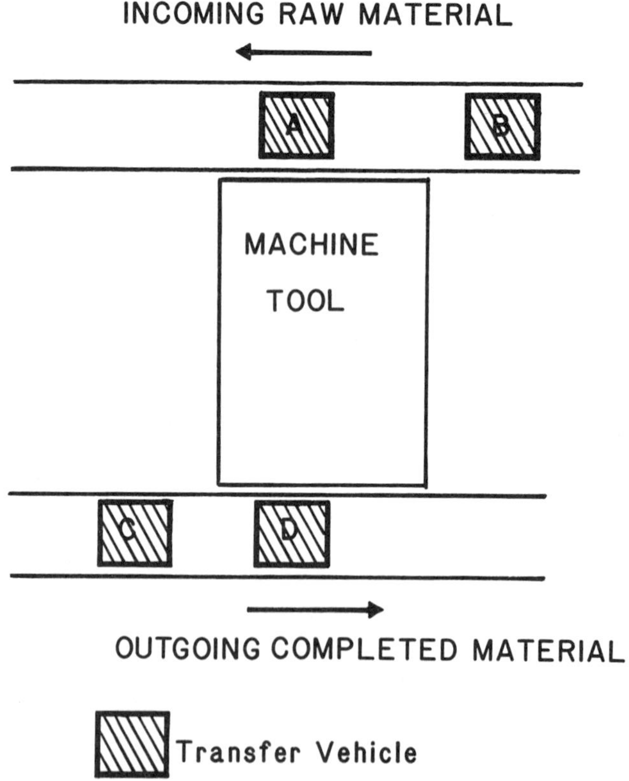

Fig. 18.1 Flexible manufacturing center with one machine tool.

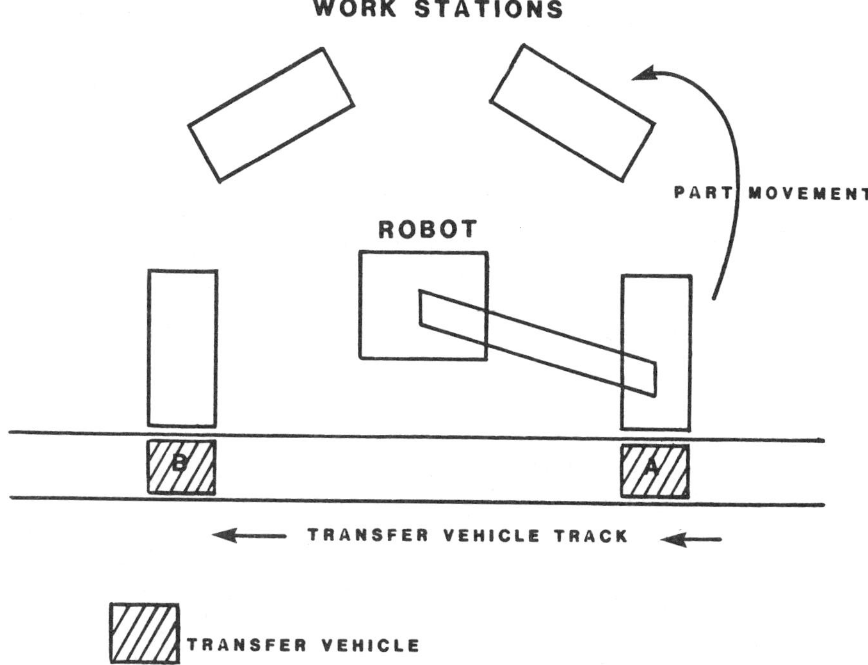

Fig. 18.2 Flexible manufacturing center with four machine tools serviced by a robot.

18.1.2 Batch Production

Studies have shown that 75% of all machined parts are produced in batches of 50 or less (1). A common example is spare parts for the aftermarket. It is not worthwhile to produce these parts in large volumes as part of the original equipment run or in large quantities after the model year has closed. The investment in inventory would be much too great for the expected annual inventory turnover in the parts. Furthermore, storage space for large quantities is a problem, and there is risk of loss through spoilage, deterioration, and obsolescence.

The traditional procedure for making a batch of small parts involves the following steps for each machine tool:

1. A work order is issued to the factory from production and inventory control. This order specifies the part name and number, the drawing number, the quantity to be made,

manufacturing information (route sheet, time standards), the raw material, the manufacturing tooling, the inspection equipment, and the required completion date.
2. The first department or machine center assembles the raw materials and tooling and puts the job on its schedule.
3. The machine is set up for the run. Tooling for the previous job is removed and the new tooling inserted into the machine. The setup is tested and cleared for production.
4. The job is run by the machinist or machine operator.
5. The material is inspected. In some plants, this is done by the machine operator. In others, the lot is moved to the inspection department, where specialized personnel do the inspecting.
6. The product is moved to the next machine center.
7. Steps #2 through #6 are repeated as many times as there are different machine tools required to finish the product.
8. The completed item is moved to storage, shipping, or assembly as needed.

This is a time-consuming process with only a small proportion of the total manufacturing time devoted to actually machining the parts. One study (2, p. 3) has shown that the average batch product spends 95% of the manufacturing cycle just waiting between centers. Only 5% of the total time is on the machine tool. While on the machine tool, the part is actually being machined only two-fifths of the time. Thus only 2% of the manufacturing cycle time goes to machining. Ninety-eight percent is spent waiting.

The major purpose and achievement of the FMS is reduction of idle time. One authority estimates that FMS raises machine utilization from 40% to 80 or 90% (2, p. 7).

18.2 ADVANTAGES OF FMS

Among the important advantages of the FMS over traditional batch manufacturing are:

1.. Reduction in manufacturing cycle time, as explained above.
2. Reduction in total inventory. A shorter cycle allows the company to reduce its inventory since it can put the lot into production in a much shorter time upon receipt of the order and since it no longer needs an inventory cushion "just in case."
3. Better tracking of inventory and work-in-process. The system can tell the location and status of each order and the current inventory situation.

4. Less "lost" inventory. Lost inventory is an item that is known to be somewhere in the factory yet is useless since its location is not known. This is a major inventory problem and expense in traditional manufacturing systems involving batch as opposed to line production.
5. Less labor required. The reduction in nonproductive machine center time means that the company is paying less for unproductive operator time. A greater proportion of direct labor goes to actual machining, a smaller share to unproductive time. This permits a reduction in the number of machine operators.
6. Better quality. Numerically or computer-controlled machine tools produce better quality with less variability. Settings and consequently dimensions can be repeated accurately. A corollary of better quality is reduction in scrap and rejects.

18.3 FLEXIBLE WORK STATION

The work station is the place where the actual operation—machining, assembly, or inspection—occurs. The hardware at this work station is similar to that used in conventional manufacturing. The major difference lies in the controls. The computer is the tool that makes the FMS possible and that eliminates time-consuming, unproductive tasks. An FMS work station may combine some manual tasks with automation.

The flexible work station consists of the following:

1. The basic machines that perform the functions of machining, assembly, and inspection.
2. Modifications such as tool changers that minimize the labor required for setups and changes.
3. Numerical control of machine tools. This reduces the need for jigs, dies, and fixtures used in traditional manufacturing.
4. Robots for material handling.
5. Automated inspection. If the part must be moved from the station for manual inspection, most of the advantages of flexible manufacturing are lost.

18.4 MATERIALS HANDLING IN THE FMS

Automated materials handling is necessary to the smooth functioning of an FMS. If the materials handling were done manually, most of the time and cost advantages of FMS would be lost. The system provides for the material to be moved to and removed from the work station on a schedule that minimizes unproductive delay between jobs.

The automated materials handling systems used in FMS are modifications of existing technology. The primary development is integrating the materials handling into the FMS. For example, a car-on-track system can be used to bring the right material to the center on time and accurately position it for the machining operation.

18.4.1 Objectives of Materials Handling in the FMS

The major objectives of the automated materials handling system in FMS are:

1. To minimize the cycle time. This is another way of saying "to increase productivity."
2. To minimize the work-in-process inventory. Since every dollar of inventory costs at least 25–30¢ a year for return on investment, obsolescence, storage, and other expenses, a reduction in inventory produces substantial savings in costs.
3. To maximize uptime. In an FMS, each element must operate reliably. If one element goes down, the whole system is nonproductive.
4. To control damage. The materials handling system must move the parts without damage or deterioration.

18.4.2 Functions of Materials Handling in FMS Operation

What's different about materials handling in an FMS from materials handling in any other system? Not much. The basic objective is still to move material to the place where it is needed, move it on time, and move it efficiently. However, an FMS poses special features and requirements that should be mentioned.

Constant or Variable Speed

Some FMS installations require variable speeds in the materials handling function. This occurs when product varies in size or other characteristics, so constant speed is undesirable. Different parts of the system may operate at different speeds to ensure efficient interfacing.

Controlled Acceleration and Deceleration

The ideal materials handling move to minimize the time of a movement is maximum acceleration to full speed at the start with maximum deceleration and a full stop just on target. However, acceleration and deceleration are controlled by physical laws. The materials handling system must be able to control the acceleration and deceleration rates to maximize efficiency and meet physical constraints. This may be done by an on-board programmable controller to relieve the system controller of this function.

Position Work Pieces Accurately and Consistently

Unlike most other materials handling systems, one in an FMS must position the work accurately within tight tolerances at the machine tool station. Limits of plus or minus a few thousandths of an inch are common. The quality of the finished machined part depends on accurate entry of the piece into the machine tool. Although this may introduce extra subroutines and complications into the materials handling system, it is necessary to the successful functioning of the system.

Queuing Capability

An FMS materials handling system should be capable of arranging product in a queue according to the order in which the product is required. For example, an order change may come through after an item is started. The system should be able to rearrange the queue to demote the lowest-priority item and move the high-priority one to the head of the queue. No matter how carefully scheduling is done, there will be changes for which queuing is the best solution.

Programmable Dispatch

The materials handling system should have the capability of dispatching the material on schedule. In an FMS, the system controller will convert the order instructions from the host computer into schedules and then issue the dispatch commands as needed.

Easy Accessibility

Accessibility to the materials handling equipment is necessary for quick maintenance or removal of a work piece that has become turned or damaged or fails to move as directed. Maintenance is particularly important. Although most FMS installations have a very good record on reliability and uptime, downtime is so expensive that quick restoration of full operation is imperative. Most FMS described in the literature have good clearances and accessibility to the subsystems and components.

Buffer Storage

The ideal FMS would have no need for buffer storage. Material would not start until called for and would proceed without delay to the machining station. Unfortunately, this isn't the way a system operates in practice. For various reasons, it may be advisable to have provision for work to enter the system ahead of the time it is required. One cause for buffering is a machine cycle much shorter than the materials handling one. Raw material may have to enter the system while the machine is working on the previous part in order for the next part to arrive at the machine station as the completed part is moving out.

Flexible-Path Patterns

In order to achieve the flexibility in an FMS, the materials handling system needs to move the material on different paths through the system. Few FMS route every item along the same path. The system must be able to move the product along the correct path for the processing required and not misdeliver or make unnecessary movements.

18.4.3 Types of Materials Handling Systems in FMS

There are several alternative materials handling systems for servicing an FMS. They include:

1. Car-on-track
2. AGV
3. Towline
4. Conveyor
5. Robots
6. Shuttles
7. Miniload automatic storage and retrieval systems
8. Carousels

These types are discussed in more detail in separate sections of this book.

18.5 COMPUTER CONTROLS

The computer that is the FMS controller is the essential characteristic which distinguishes FMS from traditional manufacturing. It controls the FMS operation through direction, feedback, and adjustment.

The FMS computer called the system controller performs the following functions:

1. Tracking
2. Materials handling control
3. Storage control
4. Work station control
5. Records maintenance

6. Performance reports
7. System simulation
8. Status reporting
9. Production scheduling
10. Communications
11. Tool control

18.5.1 Tracking

The system controller keeps track of the location and status of every work piece in the system. Without computer capabilities, this would be infeasible. Even if the function were done manually, the information could not be processed and obtained in a timely manner. Data would be too late to be useful in production decision making. The high speed of the computer enables the system to process the data, produce information in time, and deliver it to the decision maker in real time.

Tracking can serve several purposes, including:

1. Making adjustments to the production schedule to compensate for breakdowns and to meet changed priorities
2. Answering inquiries from customers and company personnel
3. Aiding scheduling of future production

Materials Handling Control

Timely movement of material to and from the flexible machining center is essential to efficient FMS operation. If materials handling were performed in the traditional manner, the time to issue the order and accomplish the move would be excessive. The system controller can direct the materials handling system to retrieve the wanted material from storage and deliver it to the work station. This can take less than a minute unless there is a queue or the system is very large. If there is a queue, this condition is communicated back to the system controller, where the decision maker can take appropriate action.

Storage Control

The FMS system controller must handle or interface with the storage system for raw material, work-in-process, and finished goods. It knows where incoming raw material or work-in-process is stored so it can direct the materials handling vehicle to retrieve the correct, unfinished product for delivery to the work station. Storage can be on a random basis; the system controller can as easily keep track of random locations as dedicated ones. A major advantage of random-access storage is that storage locations can be chosen for efficient system operation. Also, no adjustments are needed for product line changes as they are for a dedicated storage location.

Work Station Control

The machining center in an FMS is run by numerical control. The operations of the machine are controlled directly (DNC) or indirectly (NC) by a computer program. The function of the operator (if there is one) is to monitor the machine tool operation and in some installations insert and remove the part. The function of the FMS system controller is to tell the machine tool NC system what to do. This includes tooling, feeds, speeds, and dimensions. It does not get into the details of the machine tool operation; that is the function of the machine tool NC system.

Records Maintenance

It is a relatively simple matter for the system controller to convert data received into record format. Records kept include production reports, machine efficiency, and maintenance. Records can also report quality if automated inspection is built into the FMS.

Records can be produced in one of two major categories:

1. Hard copy. This is used primarily for records that must be kept for future use, such as audit or contract verification and negotiation.
2. Computer storage on disk, bubble or magnetic tape. The stored data comprise the database for current operations. Records can be displayed on a cathode-ray tube (CRT) screen if called by the operator. Since this is swift, the operator can call, interpret, and act on the data in a timely fashion. Records can be dumped when no longer needed.

Performance Reports

An important feature of the FMS is its ability to furnish management performance reports on the FMS operation. The system collects the necessary data in the course of its regular operations and stores the information in the database. It can be recalled, collated, organized, and then presented in any manner desired by management.

One advantage of performance reports prepared in this manner is that there is little opportunity or motivation to bias the reports. Every employee wants his performance presented to management in the best possible way. Under many nonautomated systems, a supervisor could bias reports. For example, he might overstate downtime to improve the productivity rate during uptime. The FMS system is so highly automated that there is little scope for biased reporting. The operations and their times are fixed by the system itself. The major and vital duty of operating personnel is to strive for maximum uptime and to fix interruptions as quickly as possible.

System Simulation

Many FMS systems have the capability to simulate operation of part or all of the system. By simulating or modeling the system mathematically, management can determine the operating characteristics of the system before it goes into operation. Modifications can be designed and tested before they are actually installed. This is substantially cheaper than a physical test on the system itself. Simulation also enables management to predict operating rates, queueing delays, maintenance downtime, and other parameters and is thus an aid to production planning and scheduling.

Status Tracking and Reporting

The FMS can provide real-time status reports on many parameters of the system. It can tell which elements are operating normally and which are not, as well as the cause of the abnormality. It can report where each part is located in the system and its current progress through the system. This information is useful for answering customer inquiries and for company supervisors and decision makers.

Status reports are usually displayed on a CRT screen rather than printed out in hard copy. This avoids several steps that add to the cost. First, most status reports are responses to specific questions. Only the information called for by the request need be displayed. This is only a small fraction of the total data collected and stored. Also, CRT display avoids the expense of making, handling, and storing hard copy if the status report is needed only for current action.

Production Scheduling

The FMS system controller can plan and schedule production according to directions for the central production scheduling system in the host computer. The production commands need only specify what is wanted and when. The FMS system controller then converts this to a specific production schedule showing machine tool work station, starting time, priority, expected completion time, routing, tooling, feeds, speeds, and dimensions. It does this by drawing on its database and on the existing schedule to optimize the FMS operation and meet the production order requirements. In this way, scheduling details are distributed away from the host computer to the FMS system controller, where it can be accomplished quicker and more efficiently.

Communications

The FMS provides communications with the personnel operating the system, those involved in related activities, management, and other elements of the computer hierarchy and distributed data-processing system. Incoming communications include queries, new data, and instructions. Outgoing communications include reports, answers to queries, and some data used elsewhere in the distributed data system.

Tool Control

A major function of an FMS is tool control. The system controller determines the tooling needed from its database, searches the system to verify its presence and location, and directs that it be sent to and inserted in the work station machine tool. It can keep track of tool wear and breakage and direct replacement when needed. Many FMS machine tool centers provide for storage of as many as 50 different cutting tools and for their insertion into and removal from the machine tool.

18.6 COMPONENTS OF FMS COMPUTER CONTROL

Most of the computer system that drives the FMS is similar to that for other materials handling systems. Computer developments affecting material handling have been discussed in Chapter 12. In this section, we shall review these developments and the components and discuss some features special to FMS.

18.6.1 Minicomputer

The heart of the FMS control system is a minicomputer. More than one may be needed, depending on the size and complexity of the system. The host computer will have the master schedule and issue appropriate production directions to the FMS minicomputer. It will collect data on inventory, shipments, and other factors but does not participate in the direct operation of the FMS. Several different makes and models of minicomputers are available for FMS.

18.6.2 Memory

FMS operation requires large quantities of data. Among the databases stored in memory are:

1. The production schedule
2. Machine tool capabilities
3. The route sheet, tooling, DNC programs for parts to be manufactured
4. Database for specifying machine feeds, speeds, and other parameters
5. The status of each machine tool and each product being worked on

6. Raw material availability
7. System performance
8. System simulation
9. Materials handling and storage
10. Fault discovery routines for key components

18.6.3 Power Supply

A major feature of the FMS computer power supply is protection against damage to and interference with the system. If the system fails to operate, the whole FMS will go down. This is especially damaging if the FMS is part of a continuous-production system which requires that all components operate for the system to produce. Also, the memory must be protected against destruction of data owing to a power supply interruption.

18.6.4 Interfaces

The FMS doesn't operate in isolation. Interfaces include the following:

1. The materials handling system, which brings raw material into the FMS and removes completed work.
2. Operator interface through CRT terminals for data editing and other functions.
3. The host computer for receiving production instructions and for reporting production, inventory, and other data.

18.6.5 Line Printers

Each FMS should include one or more line printers for generating hard-copy reports. These are essential for management use, for reporting to customers and other outsiders, for audits, and for tax review. Many protection systems call for a periodic database printout so the database can be reconstructed if a disaster wipes out data stored in memory.

18.6.6 Punched-Paper Tape

Although punched-paper tape has disappeared from most computer operations, it is still used in machine tool numerical control in many systems. It is far more tolerant of hostile environments than magnetic tape and is thus more reliable in the manufacturing shop. Punched-tape operation requires both a paper-tape punch, which need not be in the shop, and a paper-tape reader, usually at the machine tool itself.

18.6.7 Magnetic Tape Drives

Magnetic tape drives provide quick access to system data for information backup and updating.

18.6.8 Input/Output Devices

CRTs at operator work stations and other strategic points in the system provide for interaction with the control system. The operator can report status, functioning, production, and other information to the control system and can receive commands and information back from control. Other input/output devices can be used, but CRTs are by far the most common.

18.6.9 Communications Hardware

This is the wiring that transmits signals to work stations, materials handling systems, the host computer, and other interfacing systems. Communications hardware must be kept in good condition and protected from electrical, magnetic, and other interference. Static, mixed signals, and equipment failures are just a few causes of communications difficulties. Preventing failures takes good design and careful maintenance.

18.7 DESIGN AND IMPLEMENTATION

FMS design and implementation is similar in most respects to the design and implementation of any other materials handlng system discussed in Chapter 8. Here we will cover a few special features of FMS design and implementation.

Flexible manufacturing requires a major reorientation in management thinking. Managers used to conventional batch manufacturing must discard their attention to many operating decisions which will be carried out by the FMS. These functions are designed into the FMS and carried out by the FMS and should no longer receive management and supervisory attention.

Flexible manufacturing will require a new skill from managers, supervisors, engineers, and operators. Retraining present personnel or acquiring new persons with the needed skills is vital to successful FMS implementation and operation. Some of the skills needed are:

1. Real-time computer system design
2. Group technology
3. Automated materials handling system design
4. Robotics if called for by the FMS

5. Computer maintenance and repair
6. Numerical control designers and programmers
7. Production control system personnel with experience in "just-in-time" operation
8. Design engineers capable of product design for efficient FMS manufacture.

A company should introduce flexible manufacturing on a small scale. This allows time for management to become familiar with FMS and understand its place in the company and its requirements for successful operation. Time is also needed for personnel retraining in the skills required. It provides time for the company to assess the feasibility of new applications of FMS.

REFERENCES

1. J.A. Tompkins, "Flexible Manufacturing Systems," Proceedings, 1984 National Material Handling Forum, Material Handling Institute, Houston, TX, p. 3 (March 27, 1984).
2. R.F. Johoski, "Flexible Manufacturing Systems," Proceedings, 1984 National Material Handling Forum, Material Handling Institute, Houston, TX (March 28, 1984).

BIBLIOGRAPHY

Flexible Manufacturing Systems Handbook, prepared by the Charles Stark Draper Laboratory, Inc., Cambridge, MA, published by Noyes Publications, Park Ridge, NJ (1984).

19

Computers in Materials Handling and Plant Layout System Design

19.1 INTRODUCTION

The original application of digital computers in plant operations was the recording, collation, and analysis and reporting of production and cost data previously processed manually. The inventory records, the material and labor cost charges, payroll, sales, and accounts receivable were all switched from manually kept to computer records.

Within the last 5 years, computer applications have progressed further into the industrial engineering function. Analyses formerly done manually are now performed by a computer. These include engineering economy feasibility studies, risk analysis, standard time formulas, and economic lot size. This has come about through availability of cheaper computers, easier programming, easier access to the smaller dedicated computer, and familiarity with the uses and usefulness of the computer.

Computers can be used to calculate the machine loading corresponding to a given production schedule. Once the sales forecast is estimated and the operations and times required for each product determined, the computer can quickly and accurately compute the machine capacity needed to produce the product on schedule. The computer can also compute the machine loading by period and highlight future overloads and bottlenecks as well as unused capacity. Furthermore, with proper programming and periodic data update, the schedule can be adjusted for changes in sales forecasts and orders and in machine capacities (for example, extended shutdowns, major maintenance, or other causes).

A more difficult computer task is to indicate possible substitutions of one machine tool for another to balance production. Substitution is frequently not a clear-cut one-for-one process. A tool may be a satisfactory substitute for one job but be unable to meet finish or dimensional tolerances on another operation. The experienced process engineer can store much of this information in his head and use it in a complex decision process to arrive at the correct decision for substitutability.

The outputs of machine loading will be:

1. The number of hours per period scheduled for each type of machine tool. This can be extended to scheduling for each individual machine tool, but this detail inhibits supervisory ability to assign work efficiently to individual workers and machine tools.
2. The number of each type of machine tool needed to meet the production schedule. If the number on hand is insufficient, management can take appropriate action.
3. Overloads and underloads. The computer can quickly relate the scheduled production to the machine tool capacity installed and available. If there is unused capacity, the management can take steps to utilize it. If there is an overload on any machine tool type, management can take action to overcome the lack of capacity on a permanent or temporary basis.

After machine scheduling, loading, and requirements are put on the computer, the next step is to use the computer and some of the resulting data to design manufacturing and materials handling systems. The objective is a system to produce the scheduled product efficiently and cheaply. Several algorithms and computer programs have been developed to aid system design.

Another engineering innovation is computer-aided design (CAD). The engineer uses CAD to design systems, assemblies, subsystems, components, and parts. CAD draws on engineering data bases and formulas to perform the design calculations previously done laboriously by hand. In materials handling system design, CAD enables the component designer to make part and component designs that meet required specifications. With CAD, the engineer can do this cheaply, quickly, and accurately.

Algorithms and computer programs for facilities planning have been developed. The layout designer enters the space requirements, the data, and the constraints. The program then tries various combinations to reach a feasible layout which meets the requirements and constraints. It then measures the proposed layout against a predetermined criterion or set of criteria. For various reasons, these programs have been useful only to reach a first-pass feasible design proposal. They have not proved adequate to produce the best final-layout alternative. The major reasons for this inadequacy are the cost, complexity, and difficulty of obtaining data and the flexibility of the algorithms.

19.2 GENERAL DESCRIPTION

In this section, we shall describe the available computer capabilities and programs in general terms. Computer hardware is seldom a problem if the organization has good computer resources. Adequate storage for some programs may be a problem, but additional computer storage capacity is seldom difficult or expensive to acquire. The major problems are efficient algorithms, effective programs, and good data.

The time, expense, and expertise necessary to write a materials handling system design program are high. An individual company can't afford this expenditure since it doesn't have sufficient volume of system design work. Most programs have been written by a computer manufacturer, university team, or software house. The program is then leased or sold to a number of users to recover the costs of development and hopefully make a profit.

19.2.1 CAD/CAM

CAD was originally supported through government funding for defense industries. The major components are:

1. A cathode-ray tube for display (CRT)
2. An entry terminal, usually a keyboard
3. A light pen for sketching changes on the CRT image
4. CAD software for the product
5. Storage memory
6. A printer for hard-copy output

The engineer causes the engineering drawing of the item being worked on to display on the CRT. The CAD program directs the computer to make the engineering calculations using conventional engineering formulas and data in the database. If the engineering design is faulty, the engineer is notified so necessary changes can be made. The engineer then sketches proposed changes on the CRT with the light pen or enters them through the keyboard, and the computer makes the engineering calculations for the proposed changes. This process continues until the engineer is satisfied. The final design is usually printed out in hard copy for checking and other purposes. Good engineering practice calls for each engineering design to be checked by another engineer to verify the engineering accuracy of the design. The checker may use a CAD for the checking function, being careful that the checking routine doesn't duplicate the original design process in a way that perpetrates any errors that may have been made. In some installations, the hard-copy step may be omitted or minimized by sending the design directly from the CAD computer storage to manufacturing engineering and the shop floor.

CAD/CAM is good for producing detail drawings of parts and components once the system design has been finalized. At this stage, good software and design databases are available. Using CAD/CAM, the engineer or draftsman can quickly make the part and component drawings. These can be reproduced in hard copy as desired. Furthermore, changes will inevitably occur. The old design can be called up and easily modified. Also, when the design is stored in only one place, print control is easier and simpler.

19.2.2 Plant Layout Systems

Plant layout or facilities design was one of the first areas of CAD to be investigated. Since then, a number of algorithms and programs have been developed. They all follow the general procedure:

1. Data on the space requirements of departments, machines, equipment, aisles, and other space-using features are entered. This information is usually the square feet of floor space needed.
2. The constraints are entered. Examples could be prohibited location for an activity, preferred location, total floor space available, and building dimensions. Departmental dimensions and shapes may also be enetered.
3. The computer tries out various combinations of layout elements to get a complete layout that meets the constraints imposed. Iterations are run until requirements and constraints are satisfied.
4. The proposed layout is evaluated against the criteria entered previously. A common one is minimizing the distance the product moves. The output is the proposed facilities plan and its evaluation against the criteria.
5. The program may stop when a proposed layout satisfies the criteria. Some systems provide for running more than once and presenting various alternatives and their evaluations.
6. The proposed layout is printed out and used for further modification to the proposed design.

As mentioned earlier, most facilities planners have found that plant layout software can be useful for developing a starting preliminary layout, but is not adequate for producing a satisfactory final design. The computer-generated layout usually requires much added manual work to arrive at a satisfactory final layout.

19.2.3 Materials Handling Systems

Several software programs for materials handling systems design have been written. Some will carry through the whole process from problem statement and data generation to final systems design. Others will produce a good first trial from which a final design can be developed manually. Others are primarily for evaluation of proposed designs.

Few details on materials handling system design software are available publicly. Most are private proprietary programs used by a material handling equipment company to market its product. Many give good results, but must be used in connection with the owner's products.

19.2.4 Simulation

Simulation (often known as he Monte Carlo method) uses a mathematical model of the proposed materials handling system to evaluate its efficiency and cost. The key factors, such as movement rates and frequency, are entered as statistical distributions. The computer draws a value of each variable in the model at random, determines the operating results of the system according to the model and the criteria, and prints out the results. Usually, a number of trials are run to provide a large enough distribution to avoid small-sample bias and to provide the designer with better information.

Simulation is valuable for examining systems not amenable to direct mathematical analysis. Many seemingly simple systems are too complex for direct analysis. Several good simulation software programs are available, and most computer centers have one or more such programs.

19.3 ADVANTAGES

The major advantages of CAD for materials handling systems are:

1. Efficient use of designer time. Many hours can be saved in engineering computations and in alternative testing. Much of the time saved is in routine computations which require little, if any, professional skill and judgment.
2. Shorter lead time. The enhanced engineering productivity of CAD often translates into shorter lead time between

initiation of the design process and the final design. This can be worth a substantial economic return to the company.
3. A good program may prevent some errors by its check routines and checklists.

19.4 DISADVANTAGES

CAD programs have several disadvantages. Some apply to all programs; others may apply to only a few. The major ones are:

1. Inflexibility. The computer is ill equipped to make adjustments in data and parameters to suit the particular problem. Many of these adjustments are minor. An example would be reducing the width of a department area from 20 ft to 19 ft 6 in to fit available space. The computer program would say that the department doesn't fit and that it can't go in the proposed location. The human designer can look at the data and quickly decide whether the shorter distance will be satisfactory. Introducing this judgment and flexibility into the computer program is presently infeasible.
2. Data difficulties. Getting good data for a plant layout design project is difficult, even for a manually prepared design. For a computer analysis, the data task is difficult and quite costly. Again, this is due to the need for precise data for the computer while the human professional is better at using soft data which may permit the modification and adjustment.
3. Time dynamics of the system parameters. The computer requires a determinate set of data from which it derives a finite solution. However, the actual situation is rarely static. Products and product mix change, volume fluctuates, technological innovations alter parameters, and management decisions may require new data. If parameters change, the computer analysis must start all over again. Not so the human engineering professional. Changes can be more easily incorporated by the professional than by the computer.
4. Future system requirements. Related to systems changes is the need to look ahead to possible future events that may alter the system. The professional can handle this need given some insight into probable futures. Again, the computer can act only on specific data, which are generally unavailable on future trends and events.
5. Compromises. Materials handling and plant layout system design requires many compromises and accommodations among conflicting interests. These are usually worked out in face-to-face conferences between the system designers and the

officials of affected groups and departments. The computer is not capable of making these compromises. Its program follows specific and largely inflexible decision rules and does not make accommodations and compromises.

6. Complex interactions. Many materials handling system designs involve complex interactions among the components of the system and with those that interface the system. Again, the computer is ill suited for this task. Even if it could handle these complex interactions, the time and difficulty of establishing and programming these relationships would be prohibitive.

19.5 CRITERIA

Based on our discussion of advantages and disadvantages of computer programs for materials handling systems design, we can arrive at a set of criteria for evaluating the potential usefulness of a program for our purposes. These are:

1. Availability. The program must be available at reasonable cost and without major restrictions. An example of a disqualifying restriction would be prohibition of modifications to suit our needs without prior approval.
2. Compatible computer resources. Many programs are designed to run on a specific mainframe computer. Unless we have computer resources for using the program directly, the program may require extensive rewriting and modification for the company's mainframe.
3. Data. The data must be available or readily obtainable. Getting data is often the most time-consuming and consequently most expensive part of a materials handling or plant layout system design project. The program user will have to decide whether the time and expense of the program's data requirements are justified by the results expected.

   Most programs require that the input data be in a specific format. Seldom will existing data be in that format. The company should expect to collect and format the input data specifically for the program or to modify existing data to suit program needs.
4. Personnel. The company needs a team of industrial engineers and computer experts who understand the program, its uses, and its limitations. These persons should also understand factory and warehouse operations, including materials handling and plant layout.
5. Programs. The computer program should be able to do what the company wants accomplished. This calls for a clear

understanding by the vendor of the capabilities and limitations of the program. For example, a facilities planning program may produce only a first trial which will require additional manual work to arrive at an acceptable final layout. If the company expects a final product on which no further work need be done, it will be disappointed.

6. Results. Finally, the computer program should produce results that are usable by the company and are cost-effective. This calls for a clear understanding by all concerned, vendor and user both, on what can be expected from the program, what the cost will be, how long the time horizon will be from start to results, and what other resources must be committed.

19.6 GENERAL FEATURES OF COMPUTER PROGRAMS FOR MATERIALS HANDLING SYSTEM DESIGN

In this section we shall discuss some general features and subject matter of some computer programs. This is not an exhaustive listing or discussion. The reader should refer to vendor representatives or to articles and literature for more detailed information on specific programs.

19.6.1 Availability

This feature has been discussed previously, but is so important that the topic bears repetition. Anyone considering using a computer program for materials handling system or subsystem design should find out whether the program is available for his application. Many are proprietary or restricted to limited user categories. Some of the early materials handling and plant layout system design programs are no longer published.

19.6.2 Databases

A database is a body of information organized for efficient entry into, storage, and retrieval from a computer. One example is a data file of standard elemental times of manual materials handling operations. The database eliminates the need to calculate and enter each data item individually for each new application.

Good databases are available for some applications. Examples are dimensions, strength, and other engineering specifications for structural members, such as beams, channels, conveyors, and rack elements. Others use well-known engineering formulas to make necessary calculations.

Some materials handling system companies have developed systems design databases for their own products. These have the necessary data for designing the system to the customer's requirements. Information may include rates of operation, capacities, reliability, ratios, and other similar information. Most of these databases are tailored to the company needs and products and are proprietary with the manufacturer.

Other databases are nonexistent or, at best, rare. These cover such areas as space requirements for operations. Too many variables are involved to justify building adequate databases. Furthermore, much of the data will change with new technology, different market strategies, and other factors.

19.6.3 Structural Integrity

The first requirement of a materials handling system is that it be structurally sound. It must withstand the static and dynamic loads imposed and must not fail under any anticipated set of conditions.

Structural integrity may be regarded as the ability of the materials handling system to withstand loads, shocks, and accidents without damage which would interfere with the system operations. The computer analysis can be used in determining the engineering adequacy of the components and system itself.

The engineering formulas are well known and established. The computer enables the system design engineer to make and check the engineering calculations quickly and accurately. An example would be determining the strength requirements of an overhead conveyor member to meet the expected dynamic loads it will be expected to carry.

19.6.4 Performance Assessment

A major use of the computer in systems design is assessment of the system performance. The computer can use a mathematical model of the proposed system to calculate how the system will perform under various operating conditions. It can do this through direct analysis, if the system can be so modeled, or through a simulation.

Among the system performance factors that can be determined are operating rates, delay factors, and the influence of shutdowns. The operating rate is particularly important since it is the major criterion for evaluating system effectiveness in most installations.

19.6.5 Subsystem Assessment and Integration

The computer can aid in designing, evaluating, and assessing the subsystems of the overall materials handling system. Today most subsystems are standard items with the materials handling equipment manufacturer. The major problem for the user's system designer is to ensure that the subsystem is adequate for the application and that it will fit into the overall system and perform its function as needed.

19.6.6 Cost Comparisons

Cost evaluation routines are a useful feature of many systems design computer programs. Once alternative designs are determined, the computer can attach costs to each alternative from its cost database. This allows management to compare proposals in terms of costs and savings projected for each proposal.

19.6.7 Productivity Evaluation

Since the major objective of most materials handling systems is productivity improvement, measuring and evaluating the expected productivity of the proposed system is important. The computer is a good tool for this function since it can make the many calculations rapidly and accurately.

One of the more time-consuming and difficult parts of the computer productivity analysis is getting good data. Many engineers use rule-of-thumb numbers which have been established through experience. These may be sufficiently accurate for many analyses. However, the designer should review the numbers carefully to ensure that they are adequate measures of the productivity of the proposed system. If not, the designer must gather data for the specific application.

19.6.8 Graphics

Most company computer installations now have graphics capabilities. The designer can call up a graphic representation of his system on the CRT. Most computer systems can convert the graphical CRT display to hard copy.

A recent development is computer graphics that can portray the movements in a materials handling system in action. For example, automated guided vehicle (AGV) images can be moved around the guide paths on schedule to simulate the AGV moves in the actual system. The designer can use the moving graphics to check his design and to look for interferences and other potential difficulties. Another use for moving graphics is to show the proposed system and its operations to management, supervisors, and workers.

19.6.9 Implementation Schedule

Implementation schedules are an available output in computer applications to systems construction projects. They translate the data on what is required and when it is needed into schedules that show the recommended starting dates and the required completion dates for each activity going into the project. Management and other involved personnel can use the schedules to monitor progress and to take executive action to bring the project in on the scheduled completion date.

19.6.10 Total-Design Package

An ideal tool of the materials handling system designer would be a computer software package covering all aspects of the system design. Unfortunately, this is seldom practical. Among the difficulties of designing such a software package are the complexities involved and the size and number of the calculations. Another problem is making the software general enough to be widely useful. A total-design package would be unwieldy if it covered all possible alternatives. If it was a reasonable size, its use would probably be limited to an uneconomically small number of system designs.

As mentioned before, there are some system design packages for individual manufacturers. Many work well for system design using the manufacturer's products. However, they are usually proprietary and not general enough in scope or application.

19.6.11 Materials Handling Equipment Design

Several materials handling software programs are available for designing individual materials handling equipment items. These are simpler to write and use than programs for system or subsystem design. An example is a software program for designing a wooden pallet to meet the requirements of a specific application.

19.7 COMPUTER PROGRAM FOR PLANT LAYOUT

In this section, we shall discuss briefly some of the software packages for materials handling and plant layout systems design. Some are obsolete and of current interest only for what they attempted and for the reasons they are no longer available. Others are currently on the market or are used by materials handling equipment vendors for designing and evaluating systems involving their own products.

General experience has been that plant layout computer algorithms are useful in searching for alternatives and for arriving at preliminary feasible designs. They are seldom useful for fitting all the complex and indeterminate factors into a final design. A good discussion of the pros and cons of computer applications to materials handling system design appears in Collier and Footlik (1). It is difficult to standardize inputs and allow for the human element. Other negatives are lack of provision for flexibility, future adjustments, and compromises. On the other hand, computer capabilities are improving and costs decreasing relative to those of traditional engineering design work. Particularly, the improvement in computer engineering graphics and in storage and recall capabilities may shift the advantage to computer-aided materials handling system design and evaluation.

CORELAP (Computerized Relationship Layout Planning) by Lee and Moore (2) was introduced in 1967. Since then several new versions have appeared. At this writing, several improvements and modifications have been made.

ALDEP (Automated Layout Design Program) was developed by Seehof and Evans (3) of IBM and presented in 1967. It can develop an original layout and evaluate alternatives. However, IBM reports that the program is no longer available from their offices (4).

CRAFT (Computerized Relative Allocation of Facilities Technique), developed by Armour and Buffa, appeared in 1963 (5). As the first of its kind, it was widely disseminated and discussed.

Some more limited materials handling and plant layout programs are available. For example, Warden describes a microcomputer program for conveyor design (6). The problems are not complex, but the number of alternatives complicates the design process and consumes time. The program concentrates on the physical aspects of the conveyor system.

REFERENCES

1. L. M. Collier (Yes) and R. B. Footlik (No), Use of Computers in Facilities Planning, *Industrial Engineering*, Vol. 15, No. 3, pp. 50–65 (1983).

2. R. C. Lee and J. M. Moore, CORELAP—Computerized Relationship Layout Planning, *Journal of Industrial Engineering*, Vol. 18, No. 3, pp. 195–200 (1967).

3. J. M. Seehof and W. O. Evans, Automated Layout Design Program, *Journal of Industrial Engineering*, Vol. 18, No. 12, pp. 690–695 (1967).

4. Communication from the IBM Sales Office, Columbia, MO (1984).

5. G. C. Armour and E. S. Buffa, A Heuristic Algorithm and Simulation Approach to Relative Location of Facilities, *Management Science*, Vol. 9, No. 1, pp. 294–309 (1963).

6. W. H. Wardon III, Microcomputer-Aided Conveyor Design, *Industrial Engineering*, Vol. 15, No. 4, pp. 26–32 (1983).

BIBLIOGRAPHY

D. M. Anderson, New Plant Layout Information System, *Industrial Engineering*, Vol. 5, No. 4, pp. 32–37 (1973).

J. M. Apple and M. P. Deisenroth, "A Computerized Plant Layout Analysis and Evaluation Technique (PLANET)", Proceedings, American Institute of Industrial Engineers, 23rd Annual Conference and Convention, Anaheim, CA (1972).

E. S. Buffa, G. C. Armour and T. E. Vollman, Allocating Facilities with CRAFT, *Harvard Business Review*, Vol. 42, No. 2, pp. 136–159 (1964).

L. M. Collier (Yes) and R. B. Footlik (No), Use of Computers in Facilities Planning, *Industrial Engineering*, Vol. 15, No. 3, pp. 50–65 (1983).

D. H. Denhom and G. H. Brooks, "A Comparison of Three Computer Assisted Plant Layout Techniques", Proceedings, American Institute of Industrial Engineers, 21st Annual Conference and Convention, Cleveland, OH (1970).

H. Lee Hales, *Computer-Aided Facilities Planning*, Marcel Dekker, Inc., New York, (1984).

R. D. Filley, CAD for Facilities Planning: Survey Identifies Software Systems Most Useful to IEs, *Industrial Engineering*, Vol. 15, No. 3, pp. 66–80 (1983).

E. L. Fisher and S. Y. Nof, "FADES: Knowledge-Based Facility Design," 1984 Annual International Industrial Engineering Conference Proceedings, pp. 74–82.

L. R. Foulds and J. W. Giffin, "A Graph-Heuristic for Multi-Floor Building Layout", 1984 Annual International Industrial Engineering Conference Proceedings, pp. 202–205.

G. K. Gaston, Facility Layout Optimizes Space, Minimizes Costs, *Industrial Engineering*, Vol. 16, No. 5, pp. 22–27 (1984).

W. E. Goddard, Age of Simulation, *Modern Materials Handling*, Vol. 39, No. 11, p. 37 (Aug. 6, 1984).

R. V. Johnson, Spacecraft for Multi-Floor Layout Planning, *Management Science*, Vol. 28, No. 4, pp. 407–417 (1982).

B. Malakooti and G. D'Souza, "An Interactive Approach for Computer Aided Facility Layout Selection (CAFLAS)," 1984 Annual International Industrial Engineering Conference Proceedings, pp. 206–211.

S. Manivannan and D. Chaudhuri, Computer-Aided Facility Layout Algorithm Generates Alternatives to Increase Firm's Productivity, *Industrial Engineering*, Vol. 16, No. 5, pp. 81–84 (1984).

L. McGinnis, J. Trevino, and J. A. White, "A Bibliography on Material Handling Systems Analysis," Atlanta, GA, Material Handling Research Center, Georgia Institute of Technology (1983).

J. M. Moore, "Computer Aided Facilities Design: An International Survey," 2nd International Conference on Production Research (Copenhagen), Taylor & Francis, Ltd., London, England (1973).

J. M. Moore, Computer Program Evaluates Plant Layout Alternatives, *Industrial Engineering*, Vol. 3, No. 8, pp. 19–25 (1971).

R. Muther and K. McPherson, "Four Approaches to Computerized Layout Planning", *Industrial Engineering*, Vol. 2, No. 2, pp. 39–42 (1970).

T. E. Vollman and E. S. Buffa, The Facilities Layout Problem in Perspective, *Management Science*, Vol. 12, No. 10, pp. 450–468 (1966).

R. E. Webb and P. M. Wolfe, "CREATE: Interactive Model of the Building Design Process," 1984 Annual Industrial Engineering Conference Proceedings, pp. 197–201.

K. Zoller and K. Adendorff, Layout Planning by Computer Simulation, *AIIE Transactions*, Vol. 4, No. 2, pp. 116–125 (1972).

20

Organizing for Materials Handling System Operation

20.1 INTRODUCTION

Good organization of the materials handling system operation is essential for good results. The principles of good organization apply to materials handling as well as to any other function. Organization for materials handling systems design is covered in Chapter 8. In this chapter we shall look at organization for materials handling operations and for monitoring materials handling developments and performance.

The department that operates the materials handling system may be assigned some design functions. These include proposing modifications, alterations, and expansions and preparing or supporting feasibility studies. The system design project will probably require more personnel of different skills than the operating department has available. The actual design will be done by a separate design group, either internal or contract.

20.2 MATERIALS HANDLING ORGANIZATION OBJECTIVES

The first step in organizing the materials handling system operations is defining the objectives. Obviously, the first and foremost is to support production and distribution by moving the material. However, there are subsidiary objectives which support the main mission. Objectives may vary from one organization to another owing to union contracts or the importance of materials handling.

A typical set of objectives of materials handling operations organization might be:

1. To move the right material to the right destination on time
2. To minimize production downtime due to materials handling breakdowns
3. To minimize costs
4. To minimize damage and accidents
5. To supervise personnel and training effectively
6. To submit prompt and accurate reports as required

20.2.1 To Move Material

The major function and objective of the materials handling organization is to deliver the right material to the right destination at the right time. This is the reason for materials handling. Other objectives and functions are secondary. Failure to meet this objective is failure of the organization. Failure to meet other objectives in full or in part may be detrimental but is not as serious as failure to achieve the first and major objective.

20.2.2 To Minimize Downtime

In today's highly automated mass production, downtime in any part of the system will shut the whole system down. The resulting loss of production will be very costly to the company. The effectiveness of the entire production system depends on every component functioning at all times.

In some systems, it is possible by the use of work-in-process banks to buffer some operations and segments against interruption due to downtime of a component. However, this is expensive and not always satisfactory. Or there may be redundant components for downtime protection and maintenance. An example is an oil refinery which may have six parallel valves although five will do for normal functioning. The sixth is substituted for a valve that is out of service because of maintenance or accident.

20.2.3 To Minimize Costs

Materials handling operations should be carried out at minimum cost. However, management should recognize that a materials handling worker or supervisor faced with a choice between interrupting production or spending more of the company's money will spend extra money to avoid shutting down production.

20.2.4 To Minimize Damage and Accidents

Not only are accidents to material and personnel expensive, they interrupt operations and may cause a shutdown. Since materials handling accounts for a substantial proportion of all industrial accidents, the materials handling supervisor should strive to keep these to a minimum.

20.2.5 To Supervise Personnel and Training Effectively

A major part of the materials handling manager's time will be spent supervising and training workers. Achieving this objective leads to achieving the major objective of moving the material.

20.2.6 To Submit Prompt and Accurate Reports

Many materials handlers put this objective far below other objectives. Many have neither the skills, the aptitude, nor the interest to make reports and regard them as an imposition. However, it is a real objective since prompt and accurate reports are essential to management control and to efficient operation of the whole production system.

20.3 FUNCTIONS OF THE MATERIALS HANDLING ORGANIZATION

In this section, we shall discuss the functions of the materials handling organization. These closely follow the objectives listed in the previous section. Here we shall focus on what the organization and its people do rather than on what they are trying to accomplish. Again, these functions may differ from company to company owing to the nature of the industry, union contracts, industry custom, or other causes. However, these functions are necessary in every company; the organizational location may vary.

The major functions of the materials handling operating organization are as follows:

1. To move the material
2. To staff the materials handling system
3. To train materials handling personnel
4. To supervise the system
5. To report on the system's functioning
6. To report production, movement, and inventory data
7. To advise management
8. To plan

20.3.1 To Move the Material

The most important function of the materials handling organization, and the one that takes the most time and money, is moving material. This is the reason for materials handling. It is the system's prime objective.

However, the goal of the system and its supervisor should be to automate and systematize as much as feasible. The instructions, actions, and reports should be standard operating routine. Too many exceptions, rush orders, and expedited orders are symptoms of a faulty system. It should operate with a minimum of special situations and treatments. The supervisor should devote more time to planning, training, improving productivity, and investigating new technology and equipment, and less time on firefighting.

20.3.2 To Staff the Materials Handling System

A major determinant of the success of the materials handling system operation is getting and keeping good personnel. The line supervisors are particularly important since their ability influences the performance of their subordinates. Also, the materials handling manager will normally have some impact on the choice of supervisory personnel. In many plants, hourly materials handling jobs are filled primarily by seniority. This inhibits the supervisor's say in hiring, transfer, demotion, and termination of hourly employees. If the manager has carefully selected competent staff, his job will be much easier and his operation more efficient.

20.3.3 To Train Materials Handling Personnel

A major function of the materials handling supervisor is training hourly personnel. Few materials handling workers receive training in formal classes or before they start materials handling work. The first-line supervisor must instruct the workers in performance, safety, reporting, and other job elements. The line supervisor is in an ideal position for on-the-job training since he is in constant contact with the workers and can choose an appropriate moment for comments and instructions. He can also observe the workers' progress and adjust the training accordingly. There may be some formal training or workshop sessions for specialized-equipment operators, such as fork-lift truck drivers, or for safety training films.

20.3.4 To Supervise the System

No system works smoothly without direct and constant supervision. The manager must observe and keep track of personnel performance, equipment efficiency, maintenance, safety, needs of the users, and a host of other factors. Routine and emergency repairs must be planned, ordered, supervised, and checked. This doesn't mean looking into every detail and creating a "big brother" atmosphere. It does mean observing, asking questions, weighing alternative actions, and taking decisive action when required. The supervisor handles grievances and complaints.

Good materials handling supervisors are not common. Why? Many of the best are promoted to more responsible and better-paying jobs. Also, present union contracts discourage an hourly worker in a high job classification from taking a supervisory position. The pay, rewards, and status of a first-line supervisor are perceived as less desirable than those of a top-rated skilled hourly worker. The skilled worker, as a member of the bargaining unit, has no uncompensated overtime and less responsibility.

20.3.5 To Report on the System's Functioning

The system manager reports on the system's functioning to management. This reporting may be based on the "exception principle" in which only deviations from the planned norms are reported. These include breakdowns, personnel problems beyond the authority of the system manager, and recommended replacements and improvements. Some organizations require regular routine reports on the system functioning. These are usually unnecessary and are often ignored by the manager who receives them.

20.3.6 To Report Production, Movement, and Inventory Data

The materials handling system is usually the source of data on status and movement within the production system. These data are essential to smooth operation of the production system. The materials handling system is the best source of data needed by the production and inventory control department and by other departments within the organization.

The system should be organized so the data is furnished automatically with a minimum of human participation in the reporting process. In a highly automated system, reports can be sent to the system controller and the host computer on the basis of instructions received and completed or data entered by the operator into a terminal.

20.3.7 To Advise Management

The head of the materials handling department should be the person to advise management on the materials handling function. He should keep track of new technology that could improve the system. He can recommend improvements and changes. He can also advise management on future problems needing present attention. Examples would be old equipment that will need to be replaced and a demographic pattern in the work force that forecasts a number of retirements within a short period.

20.3.8 To Plan

One often neglected duty of the supervisor is planning for the future. This is not time-consuming, but does require concentrated attention and mental effort that some supervisors are unwilling to put forth. Planning requires coordination with the plans and prospects of other divisions of the company regarding future product lines, product changes, volume changes, and relocations.

20.4 PLACING THE MATERIALS HANDLING IN THE COMPANY'S ORGANIZATION STRUCTURE

Where does the materials handling organization fit into the company's overall organizational structure? Conflicting principles and objectives operate so that the place is not the same in all companies. It should go under the manager who is responsible for the production system. On the other hand, there may be good reason to put the materials handling with inventory and production control.

The final organizational decision is "Who will operate the system?" One common organizational structure is a separate materials handling department that performs all movements within the factory. This improves coordination of material functions and smoothness of the operation. The other method is to have each line supervisor responsible for moving material within his department and on to the next one. This may improve movement within the department. On the other hand, coordination of moves between departments may not be the best.

One factor in deciding organizational structure is the union contract provision concerning job descriptions, seniority lists, and job structure. This may favor a separate materials handling department if materials handling is a segregated activity under the contract. However, it will lead to inefficient use of personnel if job descriptions and functions are too restrictive.

20.5 SUGGESTED MATERIALS HANDLING ORGANIZATIONS

In this section, we shall show several examples of typical materials handling organizations and their place in the company's production system.

There are many possible ways to place the materials handling operation in the company organization. The one chosen will depend on several factors including:

1. The existing organization
2. The extent of the job
3. Availability and abilities of personnel
4. Preferences of the top management.

20.6 INTERFACES BETWEEN MATERIALS HANDLING AND OTHER DEPARTMENTS

The materials handling organization has to interface its operations with those of several other officials, departments, and functions within the company. In this section we shall talk about some of these relationships as they affect the materials handling system.

20.6.1 The Manager

The materials handling supervisor works closely with his direct supervisor in the management hierarchy in a typical manager-subordinate relationship. However, different production managers will have different personal styles. One may want to be kept informed of almost every happening in the materials handling system. Another may prefer that the materials handling system supervisor handle most matters himself and refer only major problems to the boss.

20.6.2 The Materials Handling System Designer

Company organization, policies, and customs vary greatly. Some call for close cooperation between the system designer and operating personnel. Others feel that the designer should proceed with minimal contact and "interference" from operating personnel. Best results are obtained when there is close cooperation between the designer and operating supervisor.

Some companies follow the policy of assigning operation of the materials handling system to the person who designed it. This policy motivates the designer to consider carefully the operating point of view. An example of a company that follows this path is one that uses the industrial engineering department for both materials handling system design and for supervisory training.

20.6.3 The Plant Engineer

The plant engineer and his department are assigned the responsibility for maintenance, repair, installation, and removal of equipment in most companies. It is rare for the materials handling department to provide these functions, even with its own equipment. In some companies, a few functions, such as first-echelon maintenance (checking the tires, fuel, and oil daily) or nightly battery recharging, may be done by the materials handling department.

It is important for the materials handling supervisor to handle his relationships with the plant engineering department carefully and effectively. Usually the best way is through personal contact with only the minimum of required formality and paperwork. Discussions and accommodations will usually lead to reaching mutual goals of both departments.

20.6.4 Production and Inventory Control

The production/inventory control department is the source of instructions for moving material within the plant. It is also the recipient of data from the materials handling system on production and movement data. Most of this information is automated or handled on a routine basis. However, special situations, such as rush orders, frequently come up. The two departments have to work together to accomplish the company's mission.

20.6.5 The Line Supervisor

The line supervisor is the one responsible for getting the production out. To do his job, he requires that raw material and work-in-process be delivered correctly in a timely manner. He is highly dependent on an efficient materials handling department and knows it. The goal of the materials handling system should be to give him the service he needs to do his job.

20.6.6 Receiving and Shipping

The materials handling system receives raw material from the receiving department and delivers it to storage or to a production department. The receiving department is specifically set up to receive material from outside vendors. This calls for certain functions and actions based on the vendor-vendee relationship. The materials handling system works internally and normally has little, if anything, to do with outside vendors.

Conversely, materials handling is responsible for delivering finished goods to the shipping department for movement to the customer directly or by a common carrier.

20.7 PUBLIC WAREHOUSING

A public warehouse is one that provides warehousing and related services to a number of different customers on a contract basis. There are a number of different types of public warehouses offering a wide variety of services.

One example is public warehousing in abandoned salt mines in Kansas City, Missouri. These provide plenty of space with controlled temperature and easy access. The miners dug into the side of the hill on the level and thus needed no grades or elevators for access. Construction cost was only a fraction of that for above-ground warehouses.

21

Training Materials Handling Personnel

21.1 INTRODUCTION

Contrary to public image, materials handling is not an unskilled occupation. The materials handler must have the knowledge and the skills to perform the job efficiently and to avoid accidents, injuries, and damage. This requires training in accident and damage prevention and in job skills. Training improves productivity of the resources committed to materials handling. The materials handling worker is responsible for avoiding losses due to accident and damage. These amounts may be substantial.

Training may take one of several different forms. Most important is on-the-job training by the worker's direct supervisor. Classes and training sessions are useful for some job skills. Formal training may be required for some materials handling occupations such as vehicle driver. Visual aids, brochures, and poster reminders have their place in training and in keeping workers aware of job safety and efficiency.

21.2 OBJECTIVES OF TRAINING

The first step is establishing clear objectives and guidelines for the training program. These should answer the question "What should the training accomplish?" Unless objectives are definite and clear, planning and executing a good training program is impossible. Along with the objectives, the company should set criteria for measuring progress toward the objectives and should devise a means for measuring that progress.

An example of an objective is "To operate a fork-lift truck in an efficient, safe manner." A criterion by which efficiency can be measured is pick up, move, and set down a standard load over a standard distance within a specified number of seconds. This can be checked by an actual time trial. Safe operation is harder to measure. Safe performance during a standard trial is no guarantee that the operator will handle his equipment safely over a long period under a wide variety of conditions.

21.2.1 Skills

The first major objective should be to develop the skills necessary for the worker to perform the materials handling job correctly and efficiently. How should the material be lifted onto vehicles, secured, moved, and discharged? What precautions are necessary to avoid damage to personnel, plant, equipment, and material? How can the task be done in minimum time and at minimum cost in money and effort?

One example is the correct method of lifting heavy loads manually. Incorrect lifting can result in back strain, dropped material, and excessive fatigue. Another is driving a fork-lift truck. A common assumption is that this is the same as driving an automobile. Not so. Correct lifting, moving, and setting down the material with a fork-lift truck must be learned.

21.2.2 Understanding

Today's worker is better educated than his counterpart of earlier industrial days. He or she wants to know why materials handling functions must be done in the prescribed manner and how his work fits into the overall plant operation. This understanding improves performance. In addition, the worker can often find shortcuts and improvements and be better prepared to meet emergencies.

21.2.3 Safety

A major part of the materials handling worker's job is safety. Accidents are not only painful and costly; they lose considerable time and production. An unsafe operation is seldom an efficient one. Safety training may be part of the plant's regular safety program or it may involve separate instruction for the materials handling workers.

21.2.4 Damage Control

Another important facet of the materials handling job is avoiding damage to product and equipment. This is closely related to safety in avoiding damage to people. The worker is often responsible for valuable products. Damage is an expensive, unnecessary waste; it also costs in lost production, emergency orders, and missed deliveries.

21.3 TRAINING METHODS

Different training methods and formats are available. Selecting the most effective and efficient one is important to securing desired results.

21.3.1 Formal Classes

Formal classes have several advantages. Large groups can receive the same instructional material at one presentation. The setting may be better in acoustics, lighting, and freedom from interruption. Trained instructors can be used. Coverage can be designed and controlled.

Formal classes have disadvantages too. If the class members are paid for time in class, the cost per hour can be high and the relative effectiveness low. Receptivity is often poor; a participant may be motivated by an opportunity to be off the job at full pay rather than by the opportunity to learn. The instructor may not be fully qualified.

21.3.2 On-the-Job Training

This is the most common method for training hourly workers in the United States. It has the advantage of being quick, direct, and pertinent to the job at hand. Also, on-the-job training can be varied in length, content, and approach to suit the individual worker. This reduces the amount, and therefore the cost, of unproductive training time. It can be given at the moment when it is most effective.

21.3.3 Visual Aids

Many visual aids for worker training and education are on the market. A wide range of safety posters provides constant reminders of safe practices. Movies, slide shows, and videotapes are available for instruction in materials handling and safety topics.

21.4 WHO SHOULD DO THE TRAINING?

The appropriate person to do the training is as important as the content or format. The right person can teach effectively and get the message through to the worker's attention. A poor or inappropriate instructor can be a waste of time and money.

21.4.1 Line Supervisor

Many studies have shown that the worker's immediate line supervisor is usually the most effective person to provide on-the-job training to the worker. The supervisor's position commands the worker's attention and ensures his receptivity to and complicance with the instructions. The supervisor is in day-to-day contact with the worker and can thus gauge his progress and training needs. The training can be directly and immediately administered; often a short, direct comment or instruction will replace a lecture by an outsider. The supervisor has a vested interest in the worker's progress and performance; qualified, properly trained workers will support the supervisor's performance record and make his job easier.

In one case, a transportation company wanted to improve the attitude and behavior of its employees toward passengers. A training course taught by training specialists produced negligible improvement. The course was repeated with the workers' line supervisors as instructors. The improvement was immediate and substantial.

There is a limit to the training a supervisor can impart. He has many other duties that limit his time for on-the-job training. Some training may require knowledge or skills the supervisor doesn't have.

21.4.2 Training Specialist

Some companies have a training specialist or department. This is common for large companies with frequent turnover, large resources, many employees, and specialized training needs. However, the training load tends to fluctuate; it does down or disappears when there are employees on layoff and no additions are being made to the work force. As a result, maintaining a good training staff and program over a period of years is difficult and is hard to justify in slack times.

21.4.3 Educational Instutitions

Engaging an outside educational institution may be the best way to provide necessary training. This may be most efficient if the training is extensive and requires specialized personnel and facilities. Many companies have education programs that fully or partially reimburse the employee for expenses incurred in pursuing job-related academic programs. Some educational institutions will conduct in-house training programs.

21.4.4 Senior Workers

Every new employee learns much from his senior fellow workers and associates. This training is good when the senior worker passes the positive benefits of his experience on to the newcomer. Some can be harmful, for example, output restrictions and ways to appear hard at work when little is being accomplished. Much is informal information on plant customs and supervisory personalities. This is not included in any formal instruction. Management may never realize what the worker learns from older hands.

21.5 TRAINING RECORDS AND CERTIFICATION

A record of training received should be kept for each worker. This is important in establishing the company's compliance with Occupational Health and Safety (OSHA) requirements and in determining the worker's progress and eligibility for promotion or transfer. Records need not be elaborate. They should record the type of training received, the date, the length of course, if applicable, and the worker's progress.

21.6 SAFETY TRAINING

A prime requirement of every materials handling system is safety of the worker. A substantial proportion of industrial accidents in the United States occur in materials handling activities. There are several facets to a good accident prevention program. First is engineering design to prevent accidents or to minimize the possibility of occurrence. Second is good training so the worker knows and can follow good safety practices. Third, good supervision is necessary to see that safe conditions exist and that safe practices are followed. Finally, the program requires top management support, without which no safety program can succeed.

In this section, we shall mention training in a few important safety practices. Other safety principles will be discussed in Chapter 24. For more detailed safety training, the reader is referred to specialized books and training materials. One good source is the National Safety Council (425 North Michigan Avenue, Chicago, Illinois 60611).

21.6.1 Fork-Lift Truck Operation

Safe operation of a fork-lift truck is similar to safe operation of motor vehicles. It requires training in safe operating procedures, good supervision to see that safe procedures are followed and good maintenance to ensure that the vehicles are in proper condition. In addition, the operator must be careful that the loads are properly balanced to avoid spilling and to avoid damage to the material or to the building structures. Industrial vehicles account for a substantial number of industrial accidents.

21.6.2 Lifting and Carrying

Many industrial injuries result from improper lifting and carrying of objects. The normal approach is to lean over the object and lift with the arms. This puts strain on the back and may cause injury. Bad backs are particularly troublesome because the effects may be long-term and seriously impair the worker's efficiency. Also, bad backs are notoriously difficult to diagnose and treat medically. The success rate in back operations is not high, and permanent impairment of worker efficiency often results.

The correct way to lift and carry objects puts the weight onto the legs and torso in a manner that minimizes the possibility of back damage. The worker should bend at the knees while keeping the back straight and erect. To lift, the knees are straightened with the back remaining straight and upright. Carrying manually should be limited to smaller, lighter objects. If the object is heavy, lifting and carrying equipment should be used. This equipment is described in Chapter 4.

21.6.3 Housekeeping

Poor housekeeping is a frequent cause of industrial accidents. Falls, tripping, and similar incidents account for about one-quarter of all industrial accidents. Exact statistics are unavailable, but a large proportion of these accidents are caused by objects in aisles, trash left lying around, unsafe stacking and storage, and other typical conditions resulting from poor housekeeping.

Good housekeeping requires management support and good supervision. If top management insists on good housekeeping and a clean, orderly plant, they will get it. The supervisors will take their behavioral cue from their bosses. If the first-line supervisor knows that top management wants clean housekeeping, they will see that top management gets it. Conversely, if the first-line supervisor interprets top-management actions and words to mean that housekeeping is not important, the plant will be cluttered, dirty, and full of hazards.

21.6.4 Moving Machinery

Although moving machine parts can cause serious injury, accidents from this source are among the easier ones to prevent through engineering design. Guards can be installed which prevent the operator from exposing any part of the body to moving parts. Another development has been the use of robots and automatic transfer mechanisms. These can be designed and programmed to eliminate the possibility of accident and damage. Furthermore, if an accident does occur, the result is damaged product or equipment and not an injured person. Damaged product and equipment can be replaced.

The major cause of moving-machinery accidents is failure to use the guards properly. The worker may do one of the following:

1. Remove the guard to get better production or because it is inconvenient or unpleasant
2. Not pay attention
3. Be untrained in safe procedures and precautions
4. Not recognize or report defective guards

The solution to this problem is alert supervision. The supervisor must train the workers and then see that they use the guards properly.

21.6.5 Fire and Natural Disaster

In spite of modern protection equipment, fires still occur. There are many causes, ranging from a cigarette thrown into a wastebasket full of paper to a raging chemical fire following an explosion at an oil refinery. The company's responsibilities are:

1. To make the plant as fire-safe as possible
2. To make provision for preventing injuries and saving lives by evacuation procedures
3. To call the fire department immediately
4. To minimize fire damage
5. To take steps required by the company's fire insurance policies and by fire laws and regulations

There are many building and equipment features that help prevent fires and reduce the damage from any fire that does start. Fire walls, smoke detectors, and sprinkler systems are just a few features that will improve fire protection within the plant. Many features are mandated by fire codes or by the fire insurance carrier. Others may be added by the company even though not required.

Fire protection in an older building is more difficult. Many factories were built before modern fire codes were adopted and consequently lack some fire protection features. Remodeling is usually more expensive than incorporating fire protection features into the building in the first place. It may be economical and advantageous to construct a new building in a new location. Frequently, other factors will favor constructing a new building. These include inadequate floor space, no room for expansion, antiquated multistory operations, and not enough parking or delivery space. If a new building is chosen, modern fire protection can be included at less cost than adding it to an existing structure.

The company should have procedures for orderly evacuation of the building in case of a fire alarm. Escape routes should be charted, and every employee should know his evacuation route. Fire drills should be held periodically, although this is seldom done, in the belief that it will interfere with production. In addition to evacuation plans, the company should make sure that there are adequate exits and that these exits are kept unlocked and clear at all times. The evacuation plan should also provide for alternate routes in case one or more planned routes are blocked by the fire.

The fire department should be called immediately. Fires can spread rapidly and cause considerable damage if the fire department is delayed in responding. Requiring calls to the fire department to go through certain officials may result in damaging and expensive delays. Fire department professionals much prefer to respond quickly to a number of minor fires than to be delayed reaching one important fire thay may be serious.

Before and after a fire, the company should take steps to minimize resulting damage. Of course, these should not expose personnel to injury and should not interfere with the professional fire fighters.

The contract between the company and its fire insurance carrier will impose requirements on the company. A common one is that the company shall take steps to prevent further damage after the fire is put out. An example is covering exposed material and equipment to prevent further damage from rain or snow. Another is immediate notification of the carrier's agent and cooperation with the carrier's agent and attorneys.

Other disasters may occur in various industries and in different parts of the country. Some regions are subject to tornados and windstorms. Although the area hit hard by a tornado is quite small, the force of the tornado may cause heavy damage within the limited area. The company should have evacuation procedures to prevent injury and loss of life. Usually, little can be done to prevent damage to property and equipment as the tornado hits. In some regions, special building-code requirements have been promulgated to minimize disaster damage. Examples are regulations to minimize earthquake damage in California and hurricane damage in southern Florida.

Another type of diaster is chemical spill of toxic wastes. The recent disaster in Bhopal, India, highlights the potential for death and injury from this source. If toxic materials are used, the company should do the following:

1. In design and construction, minimize the possibilities of disaster
2. In operation, devise and enforce procedures to prevent a disaster or minimize its effects
3. Install evacuation procedures in case a disaster does occur and make provision for emergency medical and other services that may be needed.

21.6.6 Safety Signs and Signals

Safety signs are used to send safety instructions to the workers and to remind the workers to observe safe practices. Safety signs include STOP signs at aisle intersections, speed limit signs, access restrictions to dangerous areas, and other similar instructions. Signs are important to safe operation of the materials handling system. However, they must be backed up by supervisory enforcement. Otherwise, they will be ignored and accidents may occur.

The most common nonvisual signals are aural ones such as horns. These are used to warn workers of approaching vehicles or moving equipment such as an overhead crane. An aural signal can penetrate the worker's concentration in a situation in which a visual signal may not.

Another type of warning signal is the flashing light. This is often used on fork-lift trucks and other industrial vehicles. The signal may be observed when an aural one is not picked up because of high noise level in the plant or impaired hearing of the worker. A combination of visual and aural signals improves the efficiency of getting the signal through. This is often a problem in a factory because of worker concentration on the job, inattention to safety hazards, or a high noise level.

BIBLIOGRAPHY

T. J. Anton, *Occupational Safety and Health Management*, McGraw-Hill, New York (1979).

J. M. Black, *Developing Competent Subordinates*, American Management Association, New York (1961).

E. Dale and L. F. Urwick, *Staff in Organization*, McGraw-Hill, New York (1960).

W. Hammer, *Occupational Safety Management and Engineering*, 2nd ed., Prentice-Hall, Englewood Cliffs, NJ (1981).

H. W. Heinrich, *Industrial Accident Prevention*, 4th ed., McGraw-Hill, New York (1959).

D. King, *Training Within the Organization*, Tavistock Publications, London (1964).

R. A. Kulwiec, Basics of Materials Handling: Training, Safety, and Maintenance, *National Safety News*, Vol. 124, No. 5, pp. 25–31 (1981).

G. Marshall, *Safety Engineering*, Brooks/Cole Engineering Division, Monterey, CA (1982).

W. McGehee and P. W. Thayer, *Training in Business and Industry*, John Wiley & Sons, New York (1961).

G. S. Ordione, *Training by Objectives*, Macmillian, New York (1970).

D. Petersen, *Techniques of Safety Management*, 2nd ed., McGraw-Hill, New York (1978).

W. P. Rodgers, *Introduction to System Safety Engineering*, John Wiley & Sons, New York (1971).

B. L. Rosenbaum, *How to Motivate Today's Workers*, McGraw-Hill, New York (1982).

M. K. Strasser, J. E. Aaron, R. C. Bohn and J. R. Eales, *Fundamentals of a Safety Education*, Macmillan, New York (1973).

Supervisor's Safety Manual, 4th ed., National Safety Council, Chicago (1973).

22

Maintenance

22.1 INTRODUCTION

An important operational aspect of a materials handling system is good maintenance of the equipment. Maintenance is defined as "to keep in an existing state (as of efficiency, repair, and validity): to preserve from failure or decline" (1). Maintenance covers a host of activities, such as repairing components and assemblies, replacing defective parts, adjusting the system for maximum operating efficiency, routine preventive actions such as lubrication, checking, and inspecting materials handling equipment, and, finally, keeping maintenance records necessary for effective work. The maintenance department may also manage the inventory of maintenance parts and supplies, although this function may be assigned to the inventory department.

A well-designed maintenance policy and organization can keep up system efficiency at reasonable cost. Unfortunately, maintenance is sometimes a top-management afterthought. Too often top management is primarily and exclusively interested in sales, production, finance, and other activities regarded as "more important." Consequently, maintenance receives inadequate attention and support. The result is increased losses from emergency mainteance and production interruptions. Also, maintenance can sometimes be deferred without immediate serious detriment to current production. Inevitably comes the day of reckoning when the high bills for previously neglected maintenance must be paid. In the meantime, the responsible executive may well have been promoted or transferred while the innocent successor is left with the blame.

Good maintenance is important to safe system operation. The designer can incorporate every possible safety feature into the system, but there are hazards against which the designer can do little. A common accident cause in materials handling is poor maintenance and housekeeping. The designer can do his best to eliminate causes such as extended body members, but he can't possibly prevent all worker actions that lead to accidents.

22.2 MAINTENANCE PRINCIPLES

22.2.1 Maintenance Objectives

The objectives of a good maintenance system are:

1. To keep the system operating at highest possible efficiency
2. To minimize lost production and downtime from system failures and breakdowns
3. To perform the maintenance function as efficiently and economically as possible

22.2.2 Functions of the Maintenance Department

The maintenance department has several duties including:

1. Repair of broken and failed parts.
2. Adjustment of parts and assemblies to full operational status.
3. Standby system operation.
4. Routine preventive maintenance such as oiling, lubrication, and data gathering.
5. Cleaning and janitorial. This may be done by a separate building custodial department.

22.2.3 Maintenance Department Organization

A maintenance department can be organized in one of several different ways. It may have subsections for each trade or occupation, or each section may be in charge of maintaining a specific section or building of the factory.

Organization may be by the degree of skill and responsibility involved. This can be by echelons or levels, as in military organizations where objectives of maximum operational readiness and best use of scarce skilled personnel and facilities are important.

First-Echelon Maintenance

This is maintenance performed by the workers operating the system, such as materials handlers, equipment operators, clerks, and others. This work includes the following:

1. Taking appropriate action when higher-level maintenance is required. This may be reporting needed work or removing the equipment from service and delivering it to maintenance.
2. Seeing that the equipment has sufficient fuel and lubricants. Checking the gas gauge, battery charge, and oil level.

Second-Echelon Maintenance

This is the maintenance performed by mechanics, electricians, and other skilled personnel in the maintenance department. It usually involves removal of defective subassemblies and replacement by new or rebuilt units. The serviceman will also check the unit for other needed repairs and make necessary adjustments to restore it to operating condition.

Third-Echelon Maintenance

This work is usually performed at a service center specializing in the type of work. An example is an engine rebuild facility. A large company may operate its own depot, but usually this is done by an outside specialized organization since few companies have the volume of work in any category to justify the expense of facilities, equipment, and personnel.

22.2.4 Preventive Maintenance

Preventive maintenace is designed to prevent future serious maintenance problems and emergencies. Small amounts of time and money are spent continuously in order to avoid large expenditures and losses later.

Preventive maintenance includes regular lubrication, minor checking, quick tests, and routine inspections. Some of these will prevent future troubles. For example, regular lubrication will prevent the major damage that comes from lack of lubricating oil on moving parts. Emergency repairs are costly by themselves, but this expense is minor compared to production interruptions. These costs are seldom identified separately in the firm's cost accounts but can be considerable.

A further function of preventive maintenance is identification of conditions calling for more extensive maintenance and repair. If a needed repair can be identified before it becomes an emergency, it can be scheduled to minimize the expense and loss from interrupted production.

22.2.5 Emergency Repairs

No matter how good the maintenance, emergencies calling for immediate maintenance action will arise. Some are due to poor preventive maintenance, some to accidents, and some to failures that could not be forecast or predicted.

The organization of maintenance for emergencies will depend on the nature of the company and its manufacturing processes. If downtime is very expensive, key maintenance personnel will be stationed near the process equipment on a standby basis. Examples are automotive assembly and steel-rolling mills. It is economical to restore production quickly even though the percent of work time of maintenance workers utilized may be very low.

In other firms the production processes are not directly interrelated. Shutdown of one machine does not bring the whole factory to a dead stop. In this case, maintenance personnel may be stationed centrally in order to maximize their utilization and effectiveness.

22.2.6 Personnel Skills

A maintenance organization needs a wide range of skills. Some of these will be furnished by in-house permanent employees. Others that are needed only occasionally may be obtained on a subcontract basis from outside firms.

A major duty of the plant engineer or maintenance supervisor is acquiring and training the maintenance personnel. He should keep track of the plant needs, the personnel available, their skills and abilities, and future needs. From this information, he should plan to ensure steady availability of needed skills. A problem faced by many companies today is an age bulge of skilled persons who will retire within a short time period. Another problem is the change in skill requirements. More computer and electronic technicians will be needed and fewer workers skilled on heavy machinery maintenance.

22.2.7 Maintenance Supervision

This is a tricky organizational problem. The direct supervisor of the maintenance worker is the foreman of the particular maintenance section. For example, the electrical foreman has charge of the electricians in maintenance work. On the other hand, the line supervisor of the production department is the one whose operations are most directly affected by the skill and efficiency of the maintenance departments. Maintenance should be organized and operated with the primary objective of providing good service to the production function. This calls for managerial and personal abilities of a higher order.

The First-Line Production Supervisor

The line supervisor is the person directly in charge of a production department. He or she is responsible for getting the production out, managing the hourly workers in his department, arranging for needed maintenance, seeing that safety and other rules and regulations are observed, and generally acting as the company's authority in his department.

In some companies the first-line supervisor is responsible for materials handling within the department and for moving completed work out to the next department. The materials handling worker reports to the first-line supervisor and is responsible to him or her. This organizational arrangement facilitates the supervisor's control over the materials handling for the best operation of the department. On the other hand, if the supervisor has too many duties and functions to look after, some may be slighted or relegated to secondary importance. Materials handling may be such a function.

The first priority of the first-line supervisor in American industry is getting out the production. He or she knows this is the principal performance criterion by which performance is judged. Whatever contributes to getting the production out will receive attention. Anything that doesn't contribute directly to output is likely to be slighted.

If the materials handling function contributes to output, the supervisor will pay attention and see that the function is carried out efficiently. If the materials handling function doesn't seem to have a direct influence on production, it may be put off. Thus, a foreman may know that certain equipment may be performing well but need maintenance in the near future. In this situation, the temptation to keep the unit in service and to postpone the maintenance is strong.

The Maintenance Supervisor

The maintenance supervisor is in charge of the maintenance of a specific area of the plant or of a specific category of work. For example, in a big factory, there may be separate maintenance supervisors for the assembly department and for the fabrication sector. This reduces the physical area over which the maintenance supervisor must operate and thus the time spent traveling around the plant. It also limits the dispersion of time, effort, and attention that can occur if the area is too big and diverse.

On the other hand, it is probably more common for each maintenance trade or specialty to have its own foreman. This enables the foreman to specialize in his skill area and keep better track of activities within his jurisdiction.

In any case, the maintenance supervisor is faced with some difficult challenges. Much maintenance work is performed in varied locations. This makes direct visual supervision of each job difficult. The work is nonstandard and difficult to measure accurately. New methods and equipment may not be easy to introduce. Among the duties of the maintenance supervisor are:

1. Scheduling work and workers
2. Responding to emergencies
3. Keeping required records
4. Discipline
5. Training
6. Handling grievances and complaints

22.2.8 Maintainability

This is the characteristic that defines the ease of keeping the system operating in good order. The major features of maintainability are designed into the system from the start. They include location of controls, lube points and panels for each access, avoiding components subject to frequent failure, and redundancy of vital components.

22.3 MAINTENANCE REQUIREMENTS OF SPECIFIC MATERIALS HANDLING EQUIPMENT

In this section, we shall look at some of the important points of maintaining specific catagories of equipment. For complete and detailed maintenance procedures and spare parts, the owner should consult the maintenance manuals of the equipment manufacturer.

22.3.1 Hand Trucks

Hand trucks require little regular maintenance. Periodic lubrication and visual inspection usually suffice. The most important factor in maintaining most materials handling equipment including hand trucks is proper use and avoiding misuse and accidents. Rough treatment will increase maintenance bills and decrease service life of any equipment. This is particularly true of some materials handling equipment which is not normally assigned to the exclusive use of a single worker who can be held responsible. If more than one person uses the hand truck, no one is responsible and consequently no one exercises great care.

22.3.2 Hand Tools and Equipment

Most hand tools require little maintenance if properly used and taken care of. Formerly, each worker was responsible for furnishing and maintaining his own hand tools and small items of equipment. Now most hand tools and small equipment items are furnished by the company. This has raised the level of maintenance and quality, although a worker may be less motivated to take good care of company tools than tools he owns and for which he is personally responsible.

Again, proper use and prompt maintenance are the key factors in good hand tool care. The worker should be trained in proper use of hand tools and small equipment. The supervisor should see that the worker uses the tools properly.

With the high cost of maintenance labor relative to the cost of small tools and hand equipment, it has become economical to discard some damaged tools and items needing more than minor maintenance. This makes sense if the repair cost is $25 an hour and a hand tool costs $5.

22.3.3 Pallets

Pallets absorb constant use and abuse which results in damage, wear, and tear. Oak pallets are the cheapest type, but are subject to broken members, splintering, or partial breaks. This requires services of a carpenter to keep the pallet supply in good working order. Someone must also judge whether the pallet is worth repairing. Oak pallets are relatively inexpensive, so it may be cheaper to discard a pallet than to repair it. They should be examined regularly to detect splinters and other damage that might cause injury to a worker even though the splinter might not interfere with the usefulness of the pallet in materials handling operations.

Some work has been done on disposable pallets. These have the advantage of requiring no maintenance since they are used only once. However, if the disposable pallet is strong enough to perform its function, it is probably too expensive in spite of the disposable feature. If the cost is kept down, the strength may not be enough to perform the pallet function.

22.3.4 Containers and Racks

Containers and racks should need little maintenance. Most will require periodic painting to avoid rust damage. If the product must be protected from contamination by oil, grease, and dirt, the racks and containers must be cleaned regularly. Another approach is to limit the use of a number of containers or pallets to the single product that could be easily contaminated. An example is sacks of flour, which must be kept free from dirt and grime.

22.3.5 Fork-Lift Trucks

Fork-lift truck maintenance requirements are similar to those for over-the-road vehicles such as automobiles and trucks. The first-echelon or daily maintenance should include:

1. Checking the oil and fuel levels and the lubrication
2. Checking tire pressure for pneumatic tires
3. Lights if needed
4. Visual check for needed repairs

Items that need to be checked on a regular basis are:

1. Battery condition for an internal combustion engine
2. Lubrication
3. Oil filters
4. Fan belts
5. Lights and electrical system

Repairs are usually made when the component fails or ceases to function at full efficiency. Examples are: engine rebuild, replacing an alternator, or a new transmission. The vehicle should also be evaluated periodically to determine whether it should be repaired to continue in service or be replaced.

22.3.6 Conveyors

Conveyors require regular lubrication. Otherwise, friction will increase the power consumption, reduce the speed, and cause excessive wear. The company should include convenient and efficient lubrication points in the original purchase specifications.

The conveyor should also be checked regularly for stretching and looseness. The drive motor and mechanisms should be inspected regularly and given preventive maintenance according to the manufacturer's instructions and company policy.

The amount and degree of preventive maintenance will also be goerned by the seriousness of conveyor downtime. If the conveyor going down will halt factory operations and cause substantial losses in production, extra care in preventive maintenance is justified. Also, special procedures to get the conveyor back in operation may be necessary. These might include stocking some spare parts near the conveyor and having mechanics available on short notice.

22.3.7 Cranes and Hoists

Cranes and hoists require regular lubrication and inspection. However, because of the accident potential from broken slings and dropped loads, extra precautions are needed. These include frequent inspection of cables and chains used to lift loads. Preventive maintenance may call for replacing parts on a regular schedule rather than waiting for failure or an inspection which discloses need for replacement. Fail-safe mechanisms should be checked regularly to minimize the possibility of an accident.

22.3.8 AS/RS and Miniload Systems

Failure of automatic storage and retrieval systems (AS/RS) and miniload systems usually results in a shutdown of part or all of the plant operations. Consequently, extra effort and preventive maintenance are warranted. An additional precaution is a standby vehicle which can replace one that is inoperative or needs repairs.

A new feature of many AS/RS and miniloads is a series of maintenance check routines built into the controlling computer system. The routine checks the components of the system and signals which, if any, components need maintenance. This is a great improvement over manual inspection since it can be done automatically by the computer with savings in labor cost. It can go through the entire checklist without skipping or ignoring one or more steps, as a human inspector might.

22.3.9 Automated Guided Vehicle Systems (AGVS)

An automated guided vehicle (AGV) is usually removed from service for maintenance. It is simple to remove an AGV needing maintenance from operations and replace it with an operating spare. The vehicle maintenance needs are similar to those of any other industrial vehicle. Lubrication, checking the batteries and electrical system, and visual inspection are much the same.

The guide paths do need attention. A painted guide path needs to be kept clear and clean. If not, the optical signals will not work and the system fails. The wire guide path needs less care and maintenance. It will keep working through many types of contaminants such as dirt and grease. The major problem is making sure that the wire is not cut inadvertently or that the circuits are not subjected to interference.

The electronic control subsystem on the AGV can be checked with a maintenance and operation computer check routine. If a defect is found, common practice is to replace the defective component with a good one and discard the defective. Examples are circuit boards and microprocessor routines.

22.3.10 Carousels

Maintenance of carousels is similar to that of other automated and semiautomated materials handling systems. Regular lubrication, periodic physical inspection of operating components, and checking the electronic and computer controls are required. Usually maintenance is performed outside regular working hours to minimize interference with operations.

22.3.11 Flexible Manufacturing Systems (FMS)

Since a flexible manufacturing system (FMS) is a complex system of interfacing subsystems, extra care is required in maintenance planning and execution. If one of the subsystems fails, it will shut down the entire FMS. This downtime will be quite costly in lost production and missed deliveries if the company uses a just-in-time inventory and production policy.

Many automatic inspection and check routines are available for FMS maintenance. These will quickly detect conditions requiring attention. The company may station maintenance workers near the FMS so action can be taken quickly. Also, spare parts and repair equipment can be stored near the system. However, as much maintenance as possible should be done in off hours. This may entail preventive replacement of key components on a regular schedule before failure occurs.

22.3.12 Automatic Identification

Maintenance of automatic identification systems falls into two categories, the automatic identification symbol with related equipment and the readers. The symbols must be clear, neat, and unambiguous. For example, if a local printer is used, care must be taken that the lines and spaces are the correct width and that small spots or voids do not cause misreads. This calls for regular cleaning and adjustment of the printers. Quality control is easier with preprinted labels which some systems use.

The readers, like other electronic gear, must be checked regularly to see that they are operating properly. Again, inspection and maintenance should be done in a way that minimizes interference with regular operations. Also, the alignment must be checked to make sure that the system gets a good read on the bar code label.

REFERENCE

1. *Webster's Ninth New Collegiate Dictionary*, Merriam-Webster, Inc., Springfield, MA, p. 718 (1983).

BIBLIOGRAPHY

F.V. Claire, Materials Handling Equipment: Preventive and Predictive Management, *National Safety News*, Vol. 122, No. 6, pp. 64–66 (1980).

T. W. Dickson, How to Conduct a Repair Shutdown, *Plant Engineering*, Vol. 37, No. 11, pp. 73–75 (May 26, 1983); Vol. 37, No. 14, pp. 58–60 (July 7, 1983); Vol. 37, No. 16, pp. 64–66 (Aug. 4, 1983).

R.A. Kulwiec, Basics of Materials Handling: Training, Safety, and Maintenance, *National Safety News*, Vol. 124, No. 5, pp. 25–31 (1981).

E.T. Newbrough, Maintenance Management Control, *Plant Engineering*, Vol. 30, No. 21, pp. 67–68 (Oct. 1982).

Service Lifts: For Fast and Safe Fork Truck Maintenance, *Modern Materials Handling*, Vol. 39, No. 9, pp. 55–57 (June 8, 1984).

23

Robots in Materials Handling

23.1 INTRODUCTION

In the last few years, industrial robots have appeared in American factories and distribution centers to perform many functions previously done by human workers. They have definite advantages for some types of materials handling applications, such as repetitive moving of heavy parts on a fixed cycle over a fixed path, materials handling in dangerous toxic and radioactive atmospheres, and taking over dangerous, dull, and boring jobs such as punch press operation. Robots also have definite limitations. They work best in a fixed-motion pattern whether it be a long or short run. Present models are deficient in vision and ability to interpret what is seen. Especially, they lack the judgment human operators possess.

The Robotic Industries Association's definition of a robot is "a reprogrammable multi-functional manipulator designed to move material, parts, tools, or specialized devices through variable programmed motions for the performance of a variety of tasks" (1). Some authorities include, in the category of robots, fixed-path manipulators, which are designed only to repeat a fixed, invariable pattern. The Japanese definition of robots is broader than ours and includes devices we don't call robots. Consequently, their robot census is higher than ours and higher than it would be if the same definition were applied in both countries.

In this section, we shall look at robots in materials handling, their characteristics, their components, their advantages and disadvantages, and some applications of robotics to materials handling.

23.2 COMPONENTS OF A ROBOT

First let's look at the separate components that make up an industrial robot and describe the function of each. Next the relationships of the components will be discussed along with planning and control.

The components of a robot are:

1. Manipulator or gripper
2. Arm
3. Power source
4. Sensors
5. Communication system
6. Image analysis
7. Control system
8. Base

23.2.1 Manipulator or Gripper

The manipulator or gripper grasps and holds the object while it is moved from origin to destination where it releases the object. The gripper must grasp the object securely, yet not mark or damage it. Some objects are heavy and difficult to hurt, whereas others are light and easily damaged. The gripper must be appropriate for the task at hand. For most applications, a standard manipulator can be used. A few applications call for a custom-designed gripper for a specialized unusual product. So far, no manipulator has the flexibility of the human hand.

Examples of manipulators are:

1. Pincer fingers, which grip the object.
2. Vacuum suction cups, which hold the object. The object must have a flat, nonporous surface for this gripper to operate properly.
3. Magnetic gripper, which holds the object by magnetic force. This can be used only with ferrous materials.

23.2.2 Arms

The arm of the robot is the part that moves the gripper with the object it holds from place to place. The arm can be elaborate, patterned after the human arm, or it can be a single, unarticulated member with only a few simple motion variations. The more elaborate arms are not as limited in their degree of freedom and motion as is a human arm. Turns of 180° are possible, allowing the robot to completely turn over an object for inspection or operation on parallel sides.

A major problem with robot arms is the weight of the arm and of the electric motors that power the motions. As much as 80% of the weight of a robot arm is in the electric motor and power systems. This places limits on the acceleration and speed of robot operation.

23.2.3 Power Sources and Motors

Power to operate the robot can be either electric or hydraulic. When robots were first introduced, many were hydraulic powered. This provides smooth operation and enough power for most applications. It also permitted the bulk of the machinery to be in the base rather than in the arms and other moving parts. However, hydraulic operation is slower, and the maintenance requirements are costlier. One estimate is that a hydraulic robot requires twice the maintenance manpower of an electric one.

In recent years electricity has powered most robots. The whole system is lighter than hydraulic. Advances in electric motors, controls, and programming have reduced problems by reducing the weight that must be carried by the robot arm. Maintenance is easier and less expensive, and downtime tends to be reduced.

23.2.4 Sensors

A sensor is a device that collects data from the environment for transmission to the robot's control and command system. It is the first step in the accomplishment of a task by a robot.

Some robots operate with only a crude, simple sensor. If the task involves picking up objects that are identical, moving them along a fixed path, and depositing the load in a known position, the need for a sensor is minimal. In fact, some robots operate with only a presence-absence sensor to verify that there is, in fact, an object in the proper position to be picked up.

Among the sensor types are:

1. Artificial vision
2. Tactile or touch
3. Proximity
4. Sonic
5. Photoelectric

23.2.5 Communication

The robot must have a communication system to send messages from sensors and other parts to the central robot controller and in turn to send messages and directions from the central controller to the manipulator, the arm, the power subsystem, and other parts of the robot.

Most communications are sent over wires from the point of message origin to its destination. A major problem is designing a communication system that will have sufficient capacity for demands placed on it. An example is vision sensing and image analysis. The number of information bits and pixels for even a simple task is astronomical. Research is underway to improve communications through such means as parallel processing, in which each of several computer subsystems processes only one part of the total information.

23.2.6 Image Analysis

Image analysis is identifying the object, its location, and its orientation based on data received from the robot's sensory system. At present, this is complex, difficult, and inefficient. An example is locating a part in a bin of jumbled parts so the robot can accurately move to pick it up.

Because of the expense and difficulties in robot vision and image analysis, good design of a robot installation calls for eliminating or minimizing the need for these functions. This can be done by designing the system so each successive part is always in the same place with the same orientation. The robot can grip the part easily since its location and position are always identical to those of previous parts.

23.2.7 Control

Control is the function of absorbing information from the environment and from the robot and then issuing directions and commands to the robot. Improvements in computers (see Chapter 12) have greatly increased the efficiency of the computers used in robotics control. Modern practice is to have a dedicated minicomputer as an integral part of the robot installation. This improves efficiency and expedites reprogramming.

Some modern robots have the flexibility to change tasks through reprogramming. This increases the versatility of the robot and allows it to be used for a series of different tasks as the needs of the factory change. For example, a robot can be reprogrammed to handle a different size and shape of part when the model changes or even to perform a completely different task.

23.3 ROBOTIC FUNCTIONS IN MATERIALS HANDLING

A robot can perforn several different types of tasks in the materials handling operations in a factory or distribution center, including:

1. Pick and place
2. Palletizing and depalletizing
3. Moving material in dangerous or toxic situations
4. Moving heavy objects that would strain or fatigue a human worker
5. Loading and unloading cars, trucks, automated guided vehicles, or other vehicles
6. Transferring parts from one materials handling subsystem to another
7. Packaging product
8. Automated inspection within a materials handling system.

23.3.1 Pick and Place

By far the most common application of robots in materials handling is the pick and place. The robot picks up an object from a fixed location, moves it to a carefully defined destination, places it in position, and then releases it. This series is repeated continuously except for interruptions due to lack of material, parts out of position, or other unusual events. The robot must be programmed to take appropriate action if any of these irregular events occur. This action may be to stop the system or to complete the cycle empty.

This type of action is suited to high-volume mass production. A robot with only three degrees of freedom in the x, y, and z axes can be used for the pick-and-place function. This reduces the capital investment in the robot and also the expense and difficulty of programming. On the other hand, converting the pick-and-place robot to another task may be a major undertaking.

23.3.2 Palletizing and Depalletizing

Another common robot application is palletizing and depalletizing. Many products cannot be handled by conventional palletizing machines as described in Chapter 4. Their size, shape, and packaging are not suited to conventional palletizers.

Palletizing is another example of a relatively simple work cycle similar to the pick and place. However, it differs since each product item has to go in a different place on the pallet. This requires a longer program but not necessarily one that is much more complicated.

The reverse of palletizing is depalletizing. A high-volume product may be received in pallet loads. Manual unloading would be expensive, tiring, and possibly entail an unacceptable damage level. The robot can quickly and efficiently unload the pallets and move the material to an inbound conveyor or to a storage location.

23.3.3 Dangerous Operations

Many industrial situations involve dangerous conditions or toxic materials. A robot can be used for these operations since it is not affected by many hazards and can be replaced in part or whole without injury to a worker if it is damaged.

For example, robots can be used in the radioactive sections of several nuclear power plants to transfer materials and fuel. The radioactivity doesn't affect the robot's performance and eliminates exposure to radioactivity. However, the robot does become radioactively contaminated once it is exposed and then must be handled as a radioactive object.

Another example of a dangerous operation is loading material into a forge or punch press. A robot can put the part into the press and remove the finished part automatically. Even if an accident to the robot occurs, replacement of the robot part is easy and no human worker is injured.

23.3.4 Moving Heavy Objects

Robots can be built to move heavy objects that weigh more than the limit considered acceptable for a human worker. If it has the capacity to handle the weight, the robot can function indefinitely without fatigue or back strain.

23.3.5 Loading and Unloading

Robots can be used for general loading and unloading if the motion patterns are within the robot's capabilities. A major limitation is the area covered and the distance the robot must move the product. Thus, a robot would have difficulty loading a railroad gondola car unless either the car or the robot were repositioned periodically while the robot filled the section within its reach.

This work is seldom sought by a manual worker. It is heavy, boring, dull, and seldom leads to a better job.

23.3.6 Packaging

A recent application is the use of robots in packaging a product.

23.3.7 Inspection

A robot can be used to move product as part of an inspection operation. The robot can pick a unit from a moving assembly line, transfer it to the inspection station, turn it over if necessary, and then put it back on the line if it is good or in the reject pile if it fails to pass inspection.

23.4 APPLICATIONS

In this section, we shall look at a few specific examples of materials handling systems that use robots. In one sense, most robots perform primarily materials handling functions. Our emphasis will be on materials handling, rather than the manufacturing or distribution operations involved.

23.4.1 Press Work

Interest in removing the operator from exposure to injury from punch presses goes back at least 30 years. At that time, some major companies adopted the policy that the press operator should not put hands into the press to feed or remove the parts. Various transfer mechanisms and guards were developed. However, many were unsatisfactory and led to reduced productivity.

Now, a robot can be used to put the material into the press and remove the completed part. Since most of the material is sheet metal, which is stiff and solid, the robot has no difficulty grasping and moving the parts. Furthermore, the robot's functions can be synchronized with the press strokes to improve productivity.

23.4.2 Forging

A robot has an additional advantage handling forging billets and parts. It is not affected as much by heat as the human operator. With properly designed grippers, the robot can pick up and carry hot forgings with no trouble or damage. It also has good repeatability, so it can accurately place the billet in each cavity. This reduces scrap losses due to misalignment.

23.4.3 Die Casting

Robots can be used to remove the completed die casting from the casting machine. Again, they are not bothered by heat.

23.4.4 Assembly

A robot can be used to put and place parts and components in an assembly operation. If the operation is high-volume mass production, the robot can often replace the human operator with resulting increase in productivity.

23.4.5 Flexible Manufacturing Systems

The role of robots in a flexible manufacturing system (FMS) is different from its function in most applications. In a typical FMS, many different parts are produced in small batches. Consequently, the robot must be programmed to handle many different items on their way through the FMS. The FMS is discussed in more detail in Chapter 18.

23.5 TRENDS

Several trends exist in materials handling by robots. Among them are:

1. Increasing gap between the cost of using a robot and that of a manual worker. The economics of scale, competition, and technological improvements are holding down the cost of robots while their efficiency goes up. On the other hand, the hourly wage increases have exceeded productivity improvement, so that unit cost of manual materials handling keeps rising.
2. Increased use of robots to load and unload machine tools. Not only do costs go down, accidents decrease.
3. The use of vision sensing will expand. Although present applications are relatively crude, research is producing improvements regularly.
4. The use of electricity for powering robots is increasing at the expense of hydraulic power.

REFERENCE

1. Robotics Industries Association, Ann Arbor, MI.

24

Safety in Materials Handling

24.1 INTRODUCTION

Safety is preventing or minimizing accidents to industrial workers. It is the absence of accidents that cause injury or harm to the individual. Since the absence of accidents is a negative concept, it is harder for the average person to conceive and to be motivated to act safely than it is for him to respond to a positive concept such as smooth materials flow.

However, safety in materials handling is vitally important. A substantial proportion of industrial accidents occur in materials handling activities. These include trips, falls, and bumps, which can incapacitate a worker for a long time.

Direct costs of accidents are substantial, yet the indirect costs are many times the direct costs. Direct costs include workmen's compensation, medical expenses, and wage payments while the worker is absent. Indirect costs are those resulting from an accident but which do not appear in an account directly identifiable as an accident cost account. Examples are time spent by supervisors and others in caring for the injured worker and the costs of lost production through disruption. Requisitioning and training a substitute is expensive, as is the impaired efficiency of the substitute. The National Safety Council estimates that the indirect costs of an accident are about nine times the direct cost.

24.2 OBJECTIVES OF A SAFETY PROGRAM

An effective safety program, like any other corporate activity, requires a clear set of goals and objectives for success. Without these goals and objectives, the safety program will flounder ineffectively.

A typical set of objectives for a safety program is:

1. To eliminate to the extent possible all avoidable hazards in the plant.
2. To minimize the possibility of accidents from causes that cannot be completely eliminated.
3. To train workers in safe practices, both general and those specific to the worker's job.
4. To create an atmosphere in which all personnel, staff, supervisory, and hourly, will work safely. To ensure that supervisors will insist on safe practices.

Other possible objectives are:

1. To reduce workmen's compensation costs to or below a certain level.
2. To have an accident record better than the industry average.
3. To set a record for the number of working days without a lost-time accident. This approach is recommended for publicity within the plant since it is easier for the average worker to understand days since the last accident than it is for him to relate to accidents per million hours worked.

24.3 MANAGEMENT'S ROLE IN SAFETY

In almost no other plant activity is management's role as critical as it is in safety. The most common single reason for failure of a safety program is lack of management interest and support. The first-level supervisors and the hourly workers will take their behavioral cues from their perception of management's interest. If they perceive that management is not really interested in safety, they will neglect it. Conversely, if they see that management means business when it says that safety is important, safety will have high priority and safe practices will be followed.

Unfortunately, management often says that safety is important when its real interests and concentration are elsewhere. In modern manufacturing, getting the production out the shipping dock is normally the highest priority. Quality is the second major interest of most managements. Unless management is careful and competent, these two goals will absorb all management's attention, and safety will suffer.

24.4 THE SAFETY PROGRAM

After management has set the goals for the company safety program and demonstrated its support for safety, a safety program must be designed and implemented. This program has several parts, all important. It requires personnel to operate it and the interest and support of the managers, supervisors, and staff.

One common situation that works against the safety program is the nature and prospects of the safety engineer's job. In some companies, this is a dead-end job with few, if any, prospects for further advancement. Consequently, safety work doesn't attract ambitious, competent persons. These people will move into line supervision or other job catagories where advancement prospects are better.

The essential components of a good safety program are similar to those of other programs. These are planning, implementation, follow-up, and periodic review. The safety program will have the following parts:

1. Prevention. Designing equipment, materials, and procedures to eliminate accident causes as much as possible.
2. Training. The workers must be trained in safe practices and attitudes. This encompasses not only safety for specific jobs, but also general safety practices such as the proper way to lift heavy objects.
3. Education. Safety education of all personnel is an ongoing, continuous activity. Posters, contests, and safety sessions can be used to impress the need for safety on the workers. Lost-time accidents occur relatively infrequently, even in dangerous industries. It's human nature to feel that an accident is improbable because one hasn't occurred for some time. Personnel need constant reminders that accidents are possible and can be serious.
4. Treatment. Unfortunately, accidents will occasionally occur even in the safest companies. The company must make provision for adequate medical treatment for the injured worker. Various levels may be required, depending on the size of the industry, the number and severity of accidents, and the availability of alternate medical service facilities.
5. Retraining and counseling. The company may have to provide for retraining the injured workers for another job.

24.5 SAFETY REQUIREMENTS OF SPECIFIC MATERIALS HANDLING EQUIPMENT

In this section, we shall look at the safety requirements and precautions for individual types of materials handling equipment. In the space available, we can only mention important points. For further material on materials handling safety, the reader should consult specialized safety writings, such as those of the National Safety Council or of the equipment manufacturer.

Although safe design and operation are essential, the key factors in accident prevention are good supervision and good training. Unfortunately, there is no way these can be written into regulations, rules, and laws and no way they could be objectively enforced were they included in regulations. Overall, it is management interest and strong support that will provide the supervision and training necessary to a good accident record.

24.5.1 Hand Trucks

Hand trucks are a frequent source of minor accidents, primarily because there are so many of them. Among the factors in safe design of hand trucks are:

1. Large enough wheels to avoid being caught in holes, troughs, and depressions in the floor
2. Hand grips of the right shape at the right height
3. Strength to avoid failure under load or impact
4. Ability to hold the load firmly without slippage

Safe design alone is not enough. Safe operation includes:

1. Proper insertation of the prongs or blade to secure the load firmly and avoid damage to the product
2. Balancing the load to avoid shifts and to position the load on the truck for easy manipulation and movement
3. Properly depositing the load to avoid tipping and damage
4. Storing unused hand trucks in a designated location or at least in an area where workers will not trip over them

24.5.2 Hand Tools and Equipment

Again, the large number of hand tools and small equipment items presents more opportunities for an accident, and consequently accidents will occur. When these tools are used in materials handling, the same safety precautions should be observed that exist for other uses of hand tools and small equipment. Among the safety principles and precautions that apply are:

1. Make sure that the tools are in good condition and do not present any hazard due to lack of tool maintenance.

2. Hand tools should not be used for any purpose other than the ones for which they were designed. For example, a screwdriver should not be used as a chiesel or pry bar.
3. The worker should be familiar with his hand tools and trained in their proper use. A person may have been using a hand tool improperly for many years simply because no training or instruction was ever provided.

24.5.3 Containers and Racks

The average person seldom thinks of a small container such as a tote box as an accident source. However, one can cause an accident. A common type is muscle or back strain through improperly lifting a loaded tote box. All workers who handle boxes, containers, and other items should receive instruction in lifting methods that avoid undue strain on the worker. Another type of accident is dropping a loaded tote box on a worker's foot. One way to minimize the chance of a dropped container is good grips; with these, the worker can hold onto the container with minimal chance of dropping the load. It is harder for a worker to hold onto an ordinary box with no provision for gripping.

Containers should be sized to keep the center of gravity close to the body and to avoid a long reach to hold the ends of the container. Keeping the container small and close to the body minimizes the chance of strain. Another reason for keeping the container small is to limit the weight of the container when loaded. There is a tendency to fill the container, even though this may exceed allowable weight limits. Some materials, such as steel, copper, and their alloys, will have a lot of weight in a small volume. It may be difficult for the worker to realize that a partially loaded container may have the maximum allowable weight.

Another consideration for small containers is stacking. If containers are to be stacked in the factory, and they usually are, they should be designed to stack securely so that there is no danger of a column of tote pans tipping over and hurting a worker. Height limits on container stacks should be set and enforced. Another reason is that lifting a loaded container above a convenient height may lead to a strain injury to the worker.

Another source of accidents from containers is slipping on a slick floor. Floors should be kept clean of oil and grease, and spills should be cleaned up promptly. Also, workers should not be expected to climb ladders carrying small containers. If a worker needs to carry something up a ladder, such as a tool or parts kit, it should be designed and carried properly to minimize accident potential.

Larger containers, such as pallet or skid loads, are designed for mechanized, not manual handling. The worker should be required to use proper equipment to lift or move large containers. Improper lifting of large or heavy loads is a frequent cause of injuries that disable the person temporarily or even permanently.

Another accident cause is lifting parts out a a partially filled skid or pallet bin. The worker may lean over and extend his arms and back so they are not centered and supported. If this movement is performed many times during the day, excessive fatigue may set in and cause an injury. Among the solutions to this problem are:

1. Use containers with lower heights. This reduces the length of reach and consequently the strain.
2. Provide the container with a drop or removable side so the worker is removing the piece directly instead of lifting it.
3. Train the worker in proper lifting technique.
4. Use dunnage or special rack fixtures to hold the parts and eliminate the bin altogether. As a layer is removed, the dunnage is taken away to uncover the next layer.

Racks and shelves should be firmly fixed to eliminate the possibility of collapsing or falling over under load or impact. A common cause of serious accidents is a vehicle collision with a rack member. Reinforcing the exposed corner members will help. Also important is monitoring vehicle traffic to emphasize careful driving. Another safety principle is neatness and good housekeeping. This eliminates falling over or running into material in the wrong place on the floor or sticking out from a rack. In addition to safety, neatness and good housekeeping promote efficiency.

A major problem in many companies is work-in-process inventory that cannot be located even though it is in the shop. Modern automated storage and retrieval (AS/RS), carousel, and miniload systems can drastically reduce the amount of lost inventory. However, automated inventory tracking does not exist in all companies nor is it universal in all companies with automated inventory tracking capabilities.

24.5.4 Fork-Lift Trucks

Fork-lift truck accidents are important not only in frequency but in severity among industrial accidents. One article describes the 10 most common types of fork-lift accidents (1):

1. Worker struck by fork-lift truck
2. Worker struck by object
3. Operator struck by falling object
4. Other worker struck by object
5. Mounting or dismounting fork-lift truck
6. Fork-lift truck overturns

7. Collision with other vehicles
8. Fork-lift truck falling off the dock
9. Body part struck by object
10. Parts falling back on the operator

In looking over the types of accidents listed above, several causes are apparent:

1. Speed too high for the operating conditions.
2. Lack of training.
3. Lack of skills.
4. Lack of driver attention to the job at hand.
5. Operator taking chances. This is often accompanied by management pressure, actual or perceived, for increased production.

Again, the above conditions may be remedied by better supervision and training.

A fork-lift operator spends most of an 8-hr day operating his vehicle. This can be fatiguing, especially if the truck is not designed in accordance with sound human engineering principles. Among those principles are:

1. Gages should be designed for quick, easy reading and should be placed where the operator can see them without straining or twisting. Only the information needed by the operator should be presented. A good principle is to have the indicator show only satisfactory or unsatisfactory rather than numerical values on a scale.
2. Levers, pedals, and other controls should be placed to reduce fatigue. They should be convenient to the operator and not require long reaches or stretches.
3. The seat should be individually adjusted to the operator and not just set for the "average" worker. The latter alignment may not suit the actual fork-lift operator and may cause extra fatigue.
4. Good visibility should be provided on all sides, especially the front.
5. Maintenance accessibility should be designed into the fork-lift truck. This is important since poor maintenance design adds to the labor time and cost with no benefit to the company.
6. Warning signals should be provided. The usual ones are audible, such as horns, and visual, such as a rotating light on top of the cab.
7. Means for easy and accurate communication between the driver and others should be provided. A related factor is the noise level, which should be kept as low as possible.

8. Overhead guards should be designed to protect the driver from falling objects.
9. Proper wheel size is important in avoiding accidents due to chuck holes and obstructions. This is particularly important in outdoor operation where smooth floors do not exist.

Pedestrian and traffic control is another way to minimize fork-lift truck accidents. If possible, fork-lift truck lanes should be exclusively for vehicles and pedestrian lanes exclusively for pedestrians. Appropriate signals and mirrors should be installed at intersections where collisions may occur.

Visual aids can play a role in safety training and indoctrination. Many good films are available from fork-lift truck manufacturers. Training courses can be arranged. In addition, the following references should be consulted:

1. American National Standards Institute (ANSI) B56 series 27 (2).
2. Occupational Health and Safety Administration general and special standards (3).
3. The manufacturers' manuals and literature.

24.5.5 Conveyors

The chief hazard from conveyors is a worker coming in contact with the moving conveyor or its load. This can be eliminated by enclosing the conveyor in such a manner that the worker cannot come in contact with any moving part. This means enclosing conveyors at worker level and installing protective screens under conveyors operating above the worker level. If the conveyor cannot be enclosed, it may be feasible to install guard rails to prevent contact.

Many conveyors can't be enclosed or guarded without interfering with the production. An example is a moving automobile assembly line. If these conveyors move at slower speeds, the opportunity for an accident and its severity are reduced. If the conveyor is open, worker carefulness and good supervision will reduce the potential for an accident.

Another safety precaution is good maintenance. Not only does this improve operational efficiency, it reduces the possibility of accident from broken parts and component failure.

Finally, measures should be taken to ensure that the product carried is secure and that none sticks out past the edge of the conveyor where it can hit a worker. A photoelectric cell to detect projecting material can be installed at moderate cost.

24.5.6 Cranes and Hoists

The potential for serious accidents from cranes and hoists is high. Most involve running into workers or dropping loads which strike a worker. The following precautions are recommended (4):

1. Make sure that the area of operation is clear of all personnel.
2. Operate the crane only over cleared areas.
3. Do not operate if the rope or chain is damaged or not properly in place
4. Make sure that the load is properly centered and secured so there is no possibility of the load tipping or falling.
5. Make sure that there are no material obstructions in the path of the crane or hoist.
6. Do not let the load sway.
7. Accelerate and move only at speeds that are not hazardous.
8. The operator must pay close attention at all times. Inattention or wandering thoughts often cause an accident.
9. Do not overload.
10. Don't leave a load suspended or unattended. When the operator leaves the crane or hoist, it should be secured in a safe position.
11. Personnel should never be lifted on a crane or hoist.

24.5.7 Automated Storage and Retrieval Systems and Miniload Systems

As with other automated systems, a major accident cause is contact with moving parts. One way to avoid accidents is to design the system to minimize worker contacts with the moving parts of the system. The system can be built so workers do not have to enter the moving system except for maintenance. This is also efficient since the computer controls that manage the automated system don't need human participation except at the order-picking and control stations and for maintenance.

Whenever maintenance must be performed, the system should be shut down and interlocked so that no movement can occur until the work is completed and shutdown interlock is removed. If possible, the maintenance should be done at night, on week-ends, or at other times when the system is not in operation. If emergency repairs are necessary while the system is in operation, it will be necessary to shut down part or all of the system while the work is being done.

The points of worker access to the system should be designed to minimize the possibility of accident or injury. This means covering moving parts such as chains and thus reducing worker exposure to moving parts. Another accident cause is falling parts. This is important since many AS/RS are high rise, thus increasing the seriousness of injury from falling objects from upper levels.

The electrical system can cause accidents if not properly protected and grounded. A system properly installed with UL-approved materials should pose little hazard as long as proper precautions are taken before the electrical components are accessed for maintenance work.

24.5.8 Automated Guided Vehicle System

In common with other vehicle systems, automated guided vehicle systems (AGVS) are subject to the usual vehicular accidents, such as collisions and loads falling off. However, there is one major difference. An AGV has a flexible bumper which stops the vehicle upon contact with a person or object. The AGV speed is so slow and the stopping action so quick that no harm is done to the person hit nor is any damage done to structures or products hit. Safety and reduced accident potential are a big plus for AGVs.

The equipment must be kept in good repair to be safe. This means checking the electrical and electronic systems regularly. Many have a built-in fault detection computer routine. If the AGV has a load and unload mechanism, this must be lubricated and maintained in good condition. Because the guide paths are either buried wire or reflective paint on the floor, there is no possibility of tripping over the guide path. Load and unload programs usually will have protective subroutines which prevent materials damage through pushing a load into an occupied space.

24.5.9 Carousels

Carousel systems are normally built in cramped quarters with little clearance between the moving carousel and the adjoining wall or other carousel. This decreases the opportunity for an accident since there is no way that the worker can come into contact with most of the carousel, whether moving or at rest. The access points are limited to one for both loading and unloading or two, one for loading and one for unloading.

Another safety feature of most carousels is a dead-man's switch for moving the carousel. The worker must keep his finger on the go button until the wanted station arrives. If he takes his hand off the control, the carousel stops. Good design of the access stations will minimize the possibility of harmful contact. Another major factor in safety is worker attention. If the picker or loader pays attention to his work, accidents are unlikely. Normally, this attention is part of the job itself.

24.5.10 Flexible Manufacturing Systems

Flexible manufacturing systems, like other automated systems, have the safety advantage of keeping the worker separate from moving parts. However, maintenance may be a safety problem. If correct procedures are adhered to, accidents from electrical shock or sudden startups will be rare.

24.6 OTHER SAFETY REQUIREMENTS

There are accident causes other than moving machinery or equipment in the materials handling function in industry. This section will discuss some of these causes and how they may be eliminated.

24.6.1 Falling and Tripping

Tripping and falls are two common types of industrial accidents in American industry. Although most accidents are minor, they can be serious, causing disability and even a fatality. The major preventive measure is good housekeeping. If floors and aisles are clear and work areas free from product and debris on the floor, tripping and falling accidents will be uncommon. Another safety measure is adequate lighting in all work areas. Many falls occur because the worker did not clearly see an obstruction in a poorly lighted section of the plant.

24.6.2 Toxic Materials

Many industries use materials that are toxic to a human worker. Chemicals, solvents, finishes, acids, and alkalies are just a few materials potentially hazardous to the worker. Among the safety precautions necessary are:

1. Segregate the storage and use of toxic materials as much as possible.
2. Clearly mark all containers of toxic materials.
3. Handle the toxic material containers in a manner that minimizes the possibility of rupture and spill.
4. Provide exhaust fans, drains, and other means of carrying away toxic wastes and spilled materials.
5. Provide protective clothing where appropriate.

6. Provide training in handling toxic materials to those who work with these materials and who may become exposed to them.
7. Provide good supervision to see that safe practices are followed.

The company should refer to the manufacturer's recommendation and to the appropriate OSHA regulations for further information on safety precautions for specific individual toxic materials.

24.6.3 Radioactive Materials

Radioactive materials pose special safety problems for protection of workers against injury and illness. Special storage units to contain the radiation are required. Handling radioactive materials must be done with mechanical manipulators in a closed compartment or with protective clothing. The radiation level is important. Some radiation is too low-level to require special precautions. Other levels are not high enough to require elaborate measures. The company should be sure that it knows what radiation hazards are present and that it takes all required and necessary precautions to avoid accidents or injuries.

24.6.4 Fire Hazards

Fire is an ever-present danger in factories and warehouses. Usually it is the product and the building that contribute fuel to the fire, and not the materials handling system or equipment. However, the materials handling system designer should be aware of the fire potential of the system he is designing and of the precautions necessary to prevent fires and to minimize fire damage if the fire occurs.

Most jurisdictions have fire codes covering such matters as sprinklers, access doors, fire hydrants, and other fire protection features. Further information may be obtained from the company's fire insurance carrier, from OSHA, or from the Underwriters Lab (UL).

A major fire protection measure often overlooked is good housekeeping. In addition to reducing the opportunities for other industrial accidents, good housekeeping reduces the possibility of fire and of damage should a fire occur.

24.6.5 Disasters

In some regions of the United States, there is potential damage from earthquake, tornado, landslide, hurricane, or other natural disaster. The materials handling system designer should be aware of these potentially destructive natural phenomena and the appropriate regulations and measures to reduce their impact should they happen.

The reader should consult applicable codes and regulations for disaster protection in his region.

REFERENCES AND NOTES

1. G. E. Lovested, Ten Top Forklift Truck Accidents, *National Safety News*, Vol. 116, No. 3, pp. 123–127 (1977).
2. American National Standards Institute, B56 Series.
3. Occupational Health and Safety Administration (OSHA) Standards are published periodically in the *Federal Register*.

   Another source is *Occupational Safety and Health Reporter*, the Bureau of National Affairs, 1231 25 Street NW, Washington, DC 20037.
4. J. W. Lahey, Overhead Safety, *National Safety News*, Vol. 122, No. 5, pp. 60–63 (1980).

BIBLIOGRAPHY

Accident Prevention Manual for Industrial Operations—Engineering and Technology, 8th ed., National Safety Council, Chicago, (1980).

L. L. Berenek, *Noise and Vibration Control*, McGraw-Hill, New York, (1971).

A. E. Florio and G. T. Stafford, *Safety Education*, 2nd ed., McGraw-Hill, New York (1962).

W. Hammer, *Occupational Safety Management and Engineering*, 2nd ed. Prentice-Hall, Englewood Cliffs, NJ (1981).

C. M. Harris, *Handbook of Noise Control*, 2nd ed. McGraw-Hill, New York (1979).

K. H. E. Kroemer, *Material Handling: Loss Control Through Ergonomics*, Alliance of American Insurers, Chicago (1979).

J. W. Lahey, Overhead Safety, *National Safety News*, Vol. 122, No. 5, pp. 20–24 (1980).

G. Marshall, *Safety Engineering*, Brooks/Cole Engineering Division, Wadsworth, Inc., Monterey, CA (1982).

Occupational Safety and Health Reporter, The Bureau of National Affairs, Washington, DC.

D. Petersen, *Techniques of Safety Management*, 2nd ed., McGraw-Hill, New York (1978).

W. P. Rodgers, *Introduction to System Safety Engineering*, John Wiley & Sons, New York (1971).

Supervisors Safety Manual, latest ed. National Safety Council, Chicago (1972).

L. F. Yerges, *Sound, Noise and Vibration Control*, 2nd ed. Van Nostrand Reinhold, New York (1978).

Index

D

E